Revise

Physics

Graham Booth

Contents

Chapter 1 Force and motion

Chapter 2 Electricity

Chapter 3 Particle physics

Chapter 4 Electromagnetic radiation

Specification lists

AQA A Physics

MODULE	SPECIFICATION TOPIC	CHAPTER REFERENCE	STUDIED IN CLASS	REVISED	PRACTICE QUESTIONS
Module 1 (M1) Particles, radiation and quantum phenomena	Particles	3.3, 3.4			
	Electromagnetic radiation and quantum phenomena	4.2, 4.4			
Module 2 (M2) Mechanics and molecular kinetic theory	Mechanics	1.1, 1.5, 1.6, 1.7, 1.8, 1.9, 1.10, 3.2			
	Molecular kinetic theory model	3.1			
Module 3 (M3) Current electricity and elastic properties of solids	Current electricity	2.1, 2.2, 2.3, 2.4			
	Elastic properties of solids	1.2, 1.3			

Examination analysis

The specification comprises three compulsory modules. The module tests consist of short structured questions, all of which are compulsory.

Module 1	1 hr test	50 marks	30%
Module 2	1 hr test	50 marks	30%
Module 3	1 hr test	50 marks	25%
	1 hr 45 min practical examination or coursework		15%

AQA B Physics

MODULE	SPECIFICATION TOPIC	CHAPTER REFERENCE	STUDIED IN CLASS	REVISED	PRACTICE QUESTIONS
Module 1 (M1) *Foundation physics*	Scalars and vectors	1.1			
	Kinematics	1.5, 1.6, 1.7, 1.10, 4.1			
	Energy concepts	1.3, 1.9			
	Electricity	2.1, 2.2, 2.3			
	d.c. circuits	2.2, 2.3			
	Information and communication				
Module 2 (M2) *Waves and nuclear physics*	Waves	4.1, 4.2			
	Diffraction and interference	4.3			
	Spectra	4.4			
	Radioactivity	3.3			
	Physics of particles	3.3, 3.4			
	Information and communication				

Examination analysis

The specification comprises three compulsory modules. The module tests consist of short answer questions and structured questions, all of which are compulsory.

Module 1	1 hr 30 min test	75 marks	35%
Module 2	1 hr 30 min test	75 marks	35%
Module 3	2 hr practical examination	78 marks	30%

Edexcel A Physics

MODULE	SPECIFICATION TOPIC	CHAPTER REFERENCE	STUDIED IN CLASS	REVISED	PRACTICE QUESTIONS
Module 1 (M1) Mechanics and radioactivity	*Rectilinear motion*	*1.5*			
	Forces and moments	*1.1, 1.2, 1.7, 1.10*			
	Dynamics	*1.6, 1.8*			
	Mechanical energy	*1.9*			
	Radioactive decay and the nuclear atom	*3.3, 3.4, 3.5*			
Module 2 (M2) Electricity and thermal physics	*Electric current and potential difference*	*2.1, 2.2*			
	Electrical circuits	*2.3*			
	Heating matter	*1.2, 3.1, 3.2*			
	Kinetic model of matter	*3.1*			
	Conservation of energy	*1.9, 3.2*			
Module 3 (M3) Topics	*A Astrophysics*				
	B Solid materials	*1.3, 1.4*			
	C Nuclear and particle physics	*3.4, 3.5*			
	D Medical physics				

Examination analysis

The specification comprises three compulsory modules. The module tests consist of structured questions, all of which are compulsory.

Module 1	*1 hr 15 min test*	*30%*
Module 2	*1 hr 15 min test*	*30%*
Module 3	*30 minute test, consisting of one question on each of the four topics. Candidates must answer the question on one topic only.*	*20%*
and	*1hr 30 min practical examination*	*20%*

Edexcel B Physics

MODULE	SPECIFICATION TOPIC	CHAPTER REFERENCE	STUDIED IN CLASS	REVISED	PRACTICE QUESTIONS
Module 1 (M1) *Physics at work, rest and play*	The sound of music	4.1, 4.2, 4.3, 4.4			
	Technology in space	1.9, 2.1, 2.2, 2.3, 3.2			
	Higher, faster, stronger	1.1, 1.5, 1.6, 1.9			
Module 2 (M2) *Physics for life*	Good enough to eat	1.2, 1.3, 1.10, 4.2			
	Digging up the past	2.1, 3.3, 4.3, 4.4			
	Spare part surgery	1.3, 1.4, 4.2, 4.4			

Examination analysis

The specification comprises three compulsory modules. The module tests consist of short-answer questions, all of which are compulsory.

Module 1	1 hr 30 min test	33.3%
Module 2	1 hr 30 min test	33.3%
Module 3	coursework	33.3%

OCR A Physics

MODULE	SPECIFICATION TOPIC	CHAPTER REFERENCE	STUDIED IN CLASS	REVISED	PRACTICE QUESTIONS
Module 1 (M1) Forces and motion	Scalars and vectors	1.1			
	Kinematics	1.5			
	Dynamics	1.1, 1.2, 1.6			
	Force, work and power	1.2, 1.4, 1.9			
	Deformation of solids	1.3			
	Forces on vehicles	1.6, 1.7, 1.9, 1.10			
	Car safety	1.6			
Module 2 (M2) Electrons and photons	Electric current	2.1, 2.2, 2.3			
	d.c. circuits	2.1, 2.2, 2.3			
	Magnetic effects of current	2.5			
	Quantum physics	4.4			
	Electromagnetic waves	4.4			
Module 3 (M3) Wave Properties	Reflection and refraction	4.2			
	Waves	4.1, 4.2			
	Superposition	4.3			

Examination analysis

The specification comprises three compulsory modules. In module tests all questions are compulsory; they consist of structured questions and questions requiring extended answers.

Module 1	1 hr test	30%
Module 2	1 hr test	30%
Module 3	45 min test	20%
and	1 hr 30 mins practical examination or coursework	20%

OCR B Physics

MODULE	SPECIFICATION TOPIC	CHAPTER REFERENCE	STUDIED IN CLASS	REVISED	PRACTICE QUESTIONS
Module 1 (M1) *Physics in action*	*Imaging and signalling*	*4.1, 4.2, 4.4*			
	Sensing	*2.1, 2.2, 2.3*			
	Designer materials	*1.2, 1.3, 1.4, 2.1, 4.2*			
Module 2 (M2) *Understanding processes*	*Waves and quantum behaviour*	*4.2, 4.3, 4.4*			
	Space, time and motion	*1.1, 1.5, 1.6, 1.9, 1.10*			

Examination analysis

The specification comprises three compulsory modules. In module tests all questions are compulsory; they consist of short focused questions and longer, structured questions with some questions being open-ended.

Module 1	1 hr 30 min test	33%
Module 2	1 hr 30 min test	37%
Module 3	coursework	30%

WJEC Physics

MODULE	SPECIFICATION TOPIC	CHAPTER REFERENCE	STUDIED IN CLASS	REVISED	PRACTICE QUESTIONS
Module 1 (M1) *Waves, light and Basics*	Basic physics	1.1, 1.2, 1.6, 1.7			
	Kinematics	1.5, 1.6, 1.10			
	Solids under stress	1.3, 1.4			
	Waves	4.1, 4.2, 4.3			
	Light	4.2, 4.3			
Module 2 (M2) *Quanta and electricity*	Conduction of electricity	2.1, 2.2, 2.3			
	Nuclear	3.3			
	Quantum physics	4.4			

Examination analysis

The specification comprises three compulsory modules. In module tests all questions are compulsory and will be 5 short answer and 2 longer questions.

Module 1	1 hr 30 min test	90 marks	35%
Module 2	1 hr 30 min test	90 marks	35%
Module 3	2 hr practical test	60 marks	30%

NICCEA Physics

MODULE	SPECIFICATION TOPIC	CHAPTER REFERENCE	STUDIED IN CLASS	REVISED	PRACTICE QUESTIONS
Module 1 (M1) Forces and electricity	Physical quantities and units	1.1			
	Static forces	1.2			
	Kinematics	1.5			
	Dynamics	1.6, 1.7, 1.8, 1.9			
	Electricity	2.1, 2.2, 2.3			
Module 2 (M2) Waves and photons	Waves	4.1, 4.2, 4.4			
	Superposition of waves	4.3			
	Photons and energy levels	4.4			
Module 3 (M3) Medical physics	Medical physics				

Examination analysis

The specification comprises three compulsory modules. The module tests consist of several compulsory short-answer questions.

Module 1	1 hr test	60 marks	33.3%
Module 2	1 hr test	60 marks	33.3%
Module 3	45 mins test	45 marks	18.4%
and	1 hr 15 mins practical examination	45 marks	15%

AS/A2 Level Physics courses

AS and A2

All Physics A Level courses are in two parts, with three separate units or modules in each part. Most students will start by studying the AS (Advanced Subsidiary) course. Some will then go on to study the second part of the A Level course, called the A2. It is also possible to study the full A Level course, both AS and A2, in any order.

How will you be tested?

Assessment units

For AS Physics, you will be tested by three assessment units. For the full A Level in Physics, you will take a further three units. AS Physics forms 50% of the assessment weighting for the full A Level.

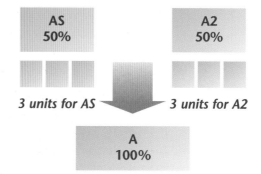

Each unit can normally be taken in either January or June. Alternatively, you can study the whole course before taking any of the unit tests. There is a lot of flexibility about when exams can be taken and the diagram below shows just some of the ways that the assessment units may be taken for AS and A Level Physics.

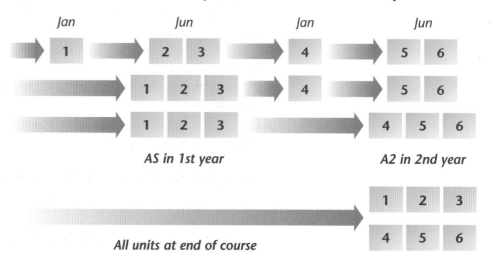

If you are disappointed with a module result, you can resit each module once. You will need to be very careful about when you take up a resit opportunity because you will have only one chance to improve your mark. The higher mark counts.

A2 and Synoptic assessment

For those students who, having studied AS, decide to go on to study A2, there are three further modules to be studied. Similar assessment arrangements apply except some module tests, those that draw together various elements of the course in a 'synoptic' assessment, have to be taken at the end of the course.

Coursework

Coursework may form part of your AS Level Physics course, depending on which specification you study. Where students have to undertake coursework, it may be a written project or for the assessment of practical skills or both. If in doubt, check with your teacher or the specification that you are studying.

Key Skills

To gain the key skills qualification, which is equivalent to an AS Level, you will need to demonstrate that you have attained level 3 in the areas of Communication, Application of number and Information technology. Part of the assessment can be done as normal class activity and part is by formal test.

What skills will I need?

For AS Physics, you will be tested by assessment objectives: these are the skills and abilities that you should have acquired by studying the course. The assessment objectives for AS Physics are shown below.

Knowledge with understanding

- recall of facts, terminology and relationships
- understanding of principles and concepts
- drawing on existing knowledge to show understanding of the responsible use of Physics in society
- selecting, organising and presenting information clearly and logically

Application of knowledge and understanding, analysis and evaluation

- explaining and interpreting principles and concepts
- interpreting and translating, from one form into another, data presented as continuous prose or in tables, diagrams and graphs
- carrying out relevant calculations
- applying knowledge and understanding to familiar and unfamiliar situations
- assessing the validity of physical information, experiments, inferences and statements

You must also present arguments and ideas clearly and logically, using specialist vocabulary where appropriate.

Experimental and investigative skills

Physics is a practical subject and part of the assessment of AS Physics will test your practical skills. This may be done during your lessons or you may be tested in a more formal practical examination. You will be assessed on four main skills:

- planning
- implementing
- analysing evidence and drawing conclusions
- evaluating evidence and procedures

Different types of questions in AS examinations

In order to assess these abilities and skills, a number of different types of question are used.

In AS Level Physics unit tests, these include short-answer questions, structured questions requiring both short answers and more extended answers, together with free-response and open-ended questions. Multiple choice question papers are not used, although it is possible that some short-answer questions will use a multiple choice format, where you have to choose the correct response from a number of given alternatives.

Short-answer questions

A short-answer question may test recall or it may test understanding by requiring you to undertake a short, one–stage calculation. Short-answer questions normally have space for the answers printed on the question paper. Here are some examples (the answers are shown in blue):

What is the relationship between electric current and charge flow?

Current = rate of flow of charge.

The current passing in a heater is 6 A when it operates from 240 V mains. Calculate the power of the heating element.

$P = I \times V = 6A \times 240V = 1440W$

Which of the following is the correct unit of acceleration?

   A m/s
   B s/m
   C m/s^2
   D m^2/s

   C m/s^2

Structured questions

Structured questions are in several parts. The parts are usually about a common context and they often become progressively more difficult and more demanding as you work your way through the question. They may start with simple recall, then test understanding of a familiar or an unfamiliar situation. The most difficult part of a structured question is usually at the end, where the candidate is sometimes asked to suggest a reason for a particular phenomenon or social implication. Most of the practice questions in this book are structured questions, as this is the main type of question used in the assessment of AS Level Physics.

When answering structured questions, do not feel that you have to complete one question before starting the next. The further you are into a question, the more difficult the marks are to obtain. If you run out of ideas, go on to the next question. Five minutes spent on the beginning of that question are likely to be much more fruitful than the same time spent wracking your brains trying to think of an explanation for an unfamiliar phenomenon.

Here is an example of a structured question that becomes progressively more demanding.

(a) A car speeds up from 20 m/s to 50 m/s in 15 s.

Calculate the acceleration of the car.

acceleration = increase in velocity ÷ time taken
= 30 m/s ÷ 15 s = 2 m/s²

(b) The total mass of the car and contents is 950 kg.

Calculate the size of the unbalanced force required to cause this acceleration.

force = mass × acceleration
= 950 kg × 2 m/s² = 1900 N

(c) Suggest why the size of the driving force acting on the car needs to be greater than the answer to (b)

The driving force also has to do work to overcome the resistive forces, e.g. air resistance and rolling resistance.

Extended answers

In AS Level Physics, questions requiring more **extended answers** will usually form part of structured questions. They will normally appear at the end of structured questions and be characterised by having at least three marks (and often more, typically five) allocated to the answers as well as several lines (up to ten) of answer space. These questions are also used to assess your abilities to communicate ideas and put together a logical argument.

The correct answers to extended questions are less well-defined than to those requiring short answers. Examiners may have a list of points for which credit is awarded up to the maximum for the question, or they may first of all judge the quality of your response as poor, satisfactory or good before allocating it a mark within a range that corresponds to that quality.

As an example of a question that requires an extended answer, a structured question on the use of solar energy could end with the following:

Suggest why very few buildings make use of solar energy in this country compared to countries in southern Europe. [5]

Points that the examiners might look for include:

- the energy from the Sun is unreliable due to cloud cover
- the intensity of the Sun's radiation is less in this country than in southern Europe due to the Earth's curvature
- more energy is absorbed by the atmosphere as the radiation has a greater depth of atmosphere to travel through
- fossil fuels are in abundant supply and relatively cheap
- the capital cost is high, giving a long payback time
- photo-voltaic cells have a low efficiency
- the energy is difficult to store for the times when it is needed the most

Full marks would be awarded for an argument that put forward three or four of these points in a clear and logical way.

Free-response questions

Little use is made of free-response and open-ended questions in AS Level Physics. These types of question allow you to choose the context and to develop your own ideas. Examples could include 'Describe a laboratory method of determining *g*, the value of free-fall acceleration' and 'Outline the evidence that suggests that light has a wave-like behaviour'. When answering this type of question it is important to plan your response and present your answer in a logical order.

Exam technique

Advanced Subsidiary Physics builds from grade CC in GCSE Science: Double Award, or equivalent in Science: Physics. This Study Guide has been written so that you will be able to tackle AS Physics from a GCSE Science background.

You should not need to search for important Physics from GCSE Science because this has been included where needed in each chapter. If you have not studied Science for some time, you should still be able to learn AS Physics using this text alone.

What are examiners looking for?

Examiners use instructions to help you to decide the length and depth of your answer.

If a question does not seem to make sense, you may have misread it – read it again!

State, define or list

This requires a short, concise answer, often recall of material that can be learnt by rote.

Explain, describe or discuss

Some reasoning or some reference to theory is required, depending on the context.

Outline

This implies a short response, almost a list of sentences or bullet points.

Predict or deduce

You are not expected to answer by recall but by making a connection between pieces of information.

Suggest

You are expected to apply your general knowledge to a 'novel' situation, one which you have not directly studied during the AS Physics course.

Calculate

This is used when a numerical answer is required. You should always use units in quantities and significant figures should be used with care.

Look to see how many significant figures have been used for quantities in the question and give your answer to this degree of accuracy.

If the question uses 3 (sig figs), then give your answer to 3 (sig figs) also.

Some dos and don'ts

Dos

Do answer the question

- No credit can be given for good Physics that is irrelevant to the question.

Do use the mark allocation to guide how much you write

- Two marks are awarded for two valid points - writing more will rarely gain more credit and could mean wasted time or even contradicting earlier valid points.

Do use diagrams, equations and tables in your responses

- Even in 'essay-type' questions, these offer an excellent way of communicating physics.

Do write legibly

- An examiner cannot give marks if the answer cannot be read.

Do write using correct spelling and grammar. Structure longer essays carefully

- Marks are now awarded for the quality of your language in exams.

Don'ts

Don't fill up any blank space on a paper

- In structured questions, the number of dotted lines should guide the length of your answer.
- If you write too much, you waste time and may not finish the exam paper. You also risk contradicting yourself.

Don't write out the question again

- This wastes time. The marks are for the answer!

Don't contradict yourself

- The examiner cannot be expected to choose which answer is intended. You could lose a hard-earned mark.

Don't spend too much time on a part that you find difficult

- You may not have enough time to complete the exam. You can always return to a difficult calculation if you have time at the end of the exam.

What grade do you want?

Everyone would like to improve their grades but you will only manage this with a lot of hard work and determination. You should have a fair idea of your natural ability and likely grade in Physics and the hints below offer advice on improving that grade.

For a Grade A

You will need to be a very good all-rounder.

- You must go into every exam knowing the work extremely well.
- You must be able to apply your knowledge to new, unfamiliar situations.
- You need to have practised many, many exam questions so that you are ready for the type of question that will appear.

The exams test all areas of the syllabus and any weaknesses in your Physics will be found out. There must be no holes in your knowledge and understanding. For a Grade A, you must be competent in all areas.

For a Grade C

You must have a reasonable grasp of Physics but you may have weaknesses in several areas and you will be unsure of some of the reasons for the Physics.

- Many Grade C candidates are just as good at answering questions as the Grade A students but holes and weaknesses often show up in just some topics.
- To improve, you will need to master your weaknesses and you must prepare thoroughly for the exam. You must become a better all-rounder.

For a Grade E

You cannot afford to miss the easy marks. Even if you find Physics difficult to understand and would be happy with a Grade E, there are plenty of questions in which you can gain marks.

- You must memorise all definitions.
- You must practise exam questions to give yourself confidence that you do know some Physics. In exams, answer the parts of questions that you know first. You must not waste time on the difficult parts. You can always go back to these later.
- The areas of Physics that you find most difficult are going to be hard to score on in exams. Even in the difficult questions, there are still marks to be gained. Show your working in calculations because credit is given for a sound method. You can always gain some marks if you get part of the way towards the solution.

What marks do you need?

As a rough guide, you will need to score an average of 40% for a Grade E, 60% for a Grade C and 80% for a Grade A:

average	80%	70%	60%	50%	40%
grade	A	B	C	D	E

Essential mathematics

This section describes some of the mathematical techniques that are needed in studying AS Level Physics.

Quantities and units

Physical quantities are described by the appropriate words or symbols, for example the symbol R is used as shorthand for the value of a *resistance*. The quantity that the word or symbol represents has both a numerical value and a unit e.g. 10.5Ω. When writing data in a table or plotting a graph, only the numerical values are entered or plotted. For this reason headings used in tables and labels on graph axes are always written as (physical quantity)/(unit), where the slash represents division. When a physical quantity is divided by its unit, the result is the numerical value of the quantity.

Resistance is an example of a **derived** quantity and the ohm is a derived unit. This means that they are defined in terms of other quantities and units. All derived quantities and units can be expressed in terms of the seven **base** quantities and units that make up the SI, or International System of Units.

The quantities, their units and symbols are shown in the table. The candela is not used in AS or A Level Physics.

quantity	unit	symbol
length	metre	m
mass	kilogram	kg
time	second	s
electric current	ampere	A
temperature difference	kelvin	K
amount of substance	mole	mol
luminous intensity	candela	cd

Equations

The equations that you use in Physics are relationships between physical quantities. Since it is not possible for one quantity to equate to a different quantity, an equation must be homogenous, i.e. the units on each side of the equation must be the same. This is useful for:

- finding the units of a constant such as resistivity
- checking the possible correctness of an equation; if the units on each side are the same the equation may be correct but if they are different it is definitely wrong.

It is important to distinguish between a physical quantity and its unit. The value of a physical quantity includes both the numerical value and the unit it is measured in.

When working out the units of an equation, make it clear that you are dealing with units only by writing units of (quantity) = (unit).

Often equations need to be used in a different form to that in which they are given or memorised. The equation needs to be **rearranged**. When rearranging an equation it is important to remember that:

- the same mathematical operation must be applied to each side of the equation
- addition or subtraction should be done first, followed by multiplication or division and finally roots and powers.

For example, to rearrange the equation $v^2 = u^2 + 2as$ to enable u to be calculated:

1 subtract $2as$ from each side of the equation to give $v^2 - 2as = u^2$

2 square root each side of the equation to give $\sqrt{(v^2 - 2as)} = u$

An equation that relates the quantity to the product or quotient of two others can be rearranged using the 'magic triangle', a triangle divided into three.

To use the triangle, start with the quantities that are multiplied or divided in the relationship. For example, in $V = IR$, you would write I and R into the triangle first. Two quantities that are multiplied together are written side-by-side at the bottom of the triangle. Where one quantity is divided by the other these are written into the triangle as they would be in a division, with one quantity 'over the other'.

The remaining quantity is now written into the vacant slot. This is illustrated below.

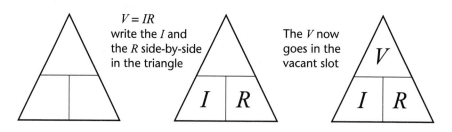

To use the triangle to make I the subject of the equation, cover up the I and you are left with 'V over R', so $I = V/R$.

Drawing graphs

Graphs have a number of uses in Physics:

- they give an immediate, visual display of the relationship between physical quantities
- they enable the values of quantities to be determined
- they can be used to confirm or disprove a hypothesis about the relationship between variables.

When plotting a graph, it is important to remember that values determined by experiment are not exact. They are subject to both the accuracy and precision of any measuring instrument and the person using it, as well as any changes that the measuring instrument itself might cause. Consider measuring the temperature of a liquid with a simple thermometer.

> Accuracy is the property of a measuring instrument. Precision refers to the reading taken; the precision depends on the user as well as the instrument being used.

- The manufacturer specifies the accuracy, the extent to which the thermometer is reliable. Typically, on a standard laboratory thermometer this is ±1°C.
- When taking a reading, the user may read to the nearest half degree or nearest degree. This describes the precision.
- Inserting the thermometer into a substance may change the temperature of the substance as the two reach thermal equilibrium.

For these reasons, having plotted experimental values on a grid, the graph line is drawn as the best straight line or smooth curve that represents the points. Where there are 'anomalous' results, i.e. points that do not fit the straight line or curve, these should always be checked. If in doubt, ignore them but do add a note in your experimental work to explain why you have ignored them and suggest how any anomalous results could have arisen.

Measurements and graphs

Quantities such as velocity, acceleration and power are defined in terms of a **rate of change** of another quantity with time. This **rate of change** can be determined by calculating the gradient of an appropriate graph. For example, *velocity* is the *rate of change of displacement with time*. Its value is represented by the gradient of a displacement-time graph. Different techniques are used to determine the gradient of a straight line and a smooth curve. For a straight line:

- determine the value of Δy, the change in the value of the quantity plotted on the y-axis, using the whole of the straight line part of the graph
- determine the corresponding value of Δx
- calculate the gradient as $\Delta y \div \Delta x$

For a smooth curve, the gradient is calculated by first drawing a tangent to the curve and then using the above method to determine the gradient of the tangent. To draw a tangent to a curve:

- mark the point on the curve where the gradient is to be determined
- use a compass to mark in two points on the curve, close to and equidistant from the point where the gradient is to be determined
- join these points with a ruler and extend the line beyond each point.

These techniques are illustrated in the diagrams below.

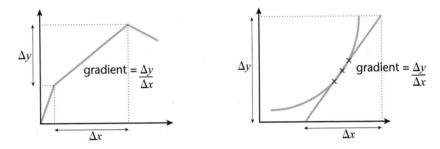

The area between a graph line and the horizontal axis, often referred to as the 'area under the graph' can also yield useful information. This is the case when the product of the quantities plotted on the axes represents another physical quantity, for example, on a *speed-time* graph this area represents the distance travelled. In the case of a straight line, the area can be calculated as that of the appropriate geometric figure. Where the graph line is curved, then the method of 'counting squares' is used.

- Count the number of complete squares between the graph line and the horizontal axis.

- Fractions of squares are counted as '1' if half the square or more is under the line, otherwise '0'.

- To work out the physical significance of each square, multiply together the quantities represented by one grid division on each axis.

These techniques are illustrated in the diagrams below

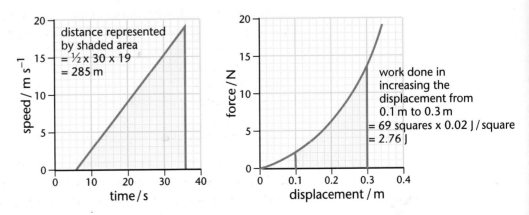

Equations from graphs

By plotting the values of two variable quantities on a suitable graph, it may be possible to determine the relationship between the variables. This is straightforward when the graph is a straight line, since all straight line graphs have an equation of the form $y = mx + c$, where m is the gradient of the graph and c is the value of y when x is zero, i.e. the intercept on the y-axis. The relationship between the variables is determined by finding the values of m and c.

The straight line graph below shows how the displacement, s, of an object varies with time, t.

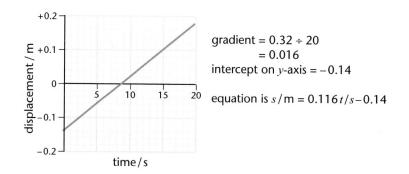

gradient = 0.32 ÷ 20
= 0.016
intercept on y-axis = −0.14
equation is $s/m = 0.116\,t/s - 0.14$

Trigonometry and Pythagoras

In the right-angled triangle shown below, the sides are labelled o (opposite), a (adjacent) and h (hypotenuse). The relationships between the size of the angle θ and the lengths of these sides are:

- $\sin \theta = o/h$
- $\cos \theta = a/h$
- $\tan \theta = o/a$

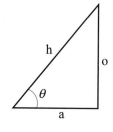

Pythagoras' theorem gives the relationship between the sides of a right-angled triangle: $h^2 = a^2 + o^2$

Multiples and submultiples

For AS Physics, you are expected to be familiar with the following multiples of units:

name	multiple	symbol
micro-	10^{-6}	μ
milli-	10^{-3}	m
kilo-	10^{3}	k
mega-	10^{6}	M
giga-	10^{9}	G

For example, the symbol MHz means 1×10^{6} Hz and mN means 1×10^{-3} N.

Equations that you need to know

This is a list of equations that may be needed but are not provided in topic tests or end-of-course examinations. Not all of these equations are needed for every specification as some are in the A Level and not the AS specification. The table shows the equations needed for each AS specification.

All other equations that are needed will be provided on a data sheet, along with any data such as the values of physical constants.

equation	symbol form	AQA A	AQA B	ED A	ED B	OCR A	OCR B	WJEC	NICCEA
speed = distance / time taken	$v = \frac{\Delta d}{t}$	•	•	•	•	•	•	•	•
force = mass × acceleration	$F = ma$	•	•	•	•	•	•	•	•
acceleration = change in velocity / time taken	$a = \frac{\Delta v}{t}$	•	•	•	•	•	•	•	•
density = mass / volume	$\rho = \frac{m}{V}$	•		•		•	•		
momentum = mass × velocity	$p = mv$	•	A2	•	A2	A2	A2	A2	•
work done = force × distance moved in direction of force	$W = Fs$	•	A2	•	•	•	•	A2	A2
power = energy transferred / time taken = work done / time taken	$P = \frac{\Delta W}{t}$	•	A2	•	•	•	•	A2	A2
weight = mass × gravitational field strength	$W = mg$	•	•	•	•	•	•	•	•
kinetic energy = ½ × mass × speed²	$E_k = \frac{1}{2} mv^2$	•	•	•	•	•	•	A2	•
change in gravitational potential energy = mass × gravitational field strength × change in height	$E_p = mgh$	•	•	•	•	•	•	A2	•
pressure = force / area	$P = \frac{F}{A}$				•		•	•	
pressure × volume = number of moles × molar gas constant × absolute temperature	$pV = nRT$	•	A2	•	A2	A2	A2	A2	A2
charge = current × time	$q = It$	•	•	•	•	•	•	•	•
potential difference = current × resistance	$V = IR$	•	•	•	•	•	•	•	•
electrical power = potential difference × current	$P = IV$	•	•	•	•	•	•	•	•
potential difference = energy transferred / charge	$V = \frac{W}{q}$	•	•	•	•	•	•	•	•
resistance = resistivity × length / cross-sectional area	$R = \frac{\rho l}{A}$	•	•	•	•	•	•	•	•
energy = potential difference × current × time	$E = VIt$	•	•	•	•	•	•	•	•
wave speed = frequency × wavelength	$v = f\lambda$	A2	•	A2	•	•	•	•	•

Four steps to successful revision

Step 1: Understand

- Study the topic to be learned slowly. Make sure you understand the logic or important concepts.
- Mark up the text if necessary – underline, highlight and make notes.
- Re-read each paragraph slowly.

GO TO STEP 2

Step 2: Summarise

- Now make your own revision note summary:
  What is the main idea, theme or concept to be learned?
  What are the main points? How does the logic develop?
  Ask questions: Why? How? What next?
- Use bullet points, mind maps, patterned notes.
- Link ideas with mnemonics, mind maps, crazy stories.
- Note the title and date of the revision notes
  (e.g. Physics: Electricity, 3rd March).
- Organise your notes carefully and keep them in a file.

This is now in **short term memory**. You will forget 80% of it if you do not go to Step 3.
GO TO STEP 3, but first take a 10 minute break.

Step 3: Memorise

- Take 25 minute learning 'bites' with 5 minute breaks.
- After each 5 minute break test yourself:
  Cover the original revision note summary
  Write down the main points
  Speak out loud (record on tape)
  Tell someone else
  Repeat many times.

The material is well on its way to **long term memory**.
You will forget 40% if you do not do step 4. **GO TO STEP 4**

Step 4: Track/Review

- Create a Revision Diary (one A4 page per day)
- Make a revision plan for the topic, e.g. 1 day later, 1 week later, 1 month later.
- Record your revision in your Revision Diary, e.g.
  Physics: Electricity, 3rd March 25 minutes
  Physics: Electricity, 5th March 15 minutes
  Physics: Electricity, 3rd April 15 minutes
  ... and then at monthly intervals.

Chapter 1

Force and motion

The following topics are covered in this chapter:

- *Forces and vectors*
- *Some effects of forces*
- *Under stress*
- *Solid materials*
- *Motion in one and two dimensions*

- *Force and acceleration*
- *Newton I and III*
- *Momentum and the second law*
- *Energy to work*
- *Drag and terminal velocity*

1.1 Forces and vectors

After studying this section you should be able to:

- *recognise and describe forces*
- *distinguish between a vector and a scalar quantity*
- *add together two vectors and subtract one vector from another*
- *split a vector into two parts at right angles to each other*

LEARNING SUMMARY

Describing forces

AQA A	M2	OCR A	M1
EDEXCEL A	M1	OCR B	M2
EDEXCEL B	M1	WJEC	M1
NICCEA	M1		

In this context 'normal' means 'at right angles to the surface'.

The first bullet point here is emphasising that all forces are caused by objects and they act on other objects.

Forces can have different effects; they can cause objects to start moving, stop moving, stay in the same place or rotate. They can also be of different types; **gravitational** forces are caused by and affect objects with mass, **friction** forces oppose relative motion and the force pushing up on you at the moment, the **normal contact** force, is due to compression of the material that you are sat on.

Forces also have things in common:

- all forces can be described as **object A pulls/pushes object B**
- all forces can be represented in both size and direction by an arrow on a diagram.

Here are some examples.

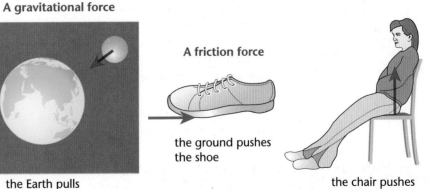

A gravitational force

A friction force

A normal contact force

the Earth pulls
the moon

the ground pushes
the shoe

the chair pushes
the person

Drawing the arrows

AQA A	M2	OCR A	M1
EDEXCEL A	M1	OCR B	M2
EDEXCEL B	M1	WJEC	M1
NICCEA	M1		

Of the three examples given above, two of the forces are due to contact between the objects. In each case the arrow representing the force has been drawn in the centre of the area of contact. Gravitational forces act at a distance with no contact being necessary. These forces are always drawn as if they act at the centre of gravity, or centre of mass, of the object.

the centre of mass of a traffic cone, a rubber ring and a person

The gravitational force acting on an object due to the planet or moon whose surface it is on is known as its weight.

> **Weight is related to mass by the formula:**
> weight = mass × gravitational field strength
> $W = m \times g$
>
> **KEY POINT**

Physical quantities that have direction as well as size are called **vectors**. Quantities with size only are **scalars**. Some examples are given in the table.

vectors	scalars
force	mass
velocity	speed
acceleration	length
displacement	distance
field strength	energy

To add together two scalar quantities the normal rules of arithmetic apply, for example, 2 kg + 3 kg = 5 kg and no other answers are possible. When adding vector quantities, both the size and direction have to be taken into account.

What is the sum of a 2 N force and a 3 N force acting on the same object? The answer could be any value between 1 N and 5 N, depending on the directions involved.

Gravitational field strength and free fall acceleration are the same physical quantity. Close to the Earth's surface, g has the value 9.8 N kg⁻¹ or 9.8 m s⁻², In calculations, the value of 10 m s⁻² is often used as a good approximation.

When representing a vector quantity on a diagram, an arrow is always used to show its direction.

The sum of two vectors

AQA A	M2	NICCEA	M1
AQA B	M1	OCR A	M1
EDEXCEL A	M1	OCR B	M2
EDEXCEL B	M1	WJEC	M1

The sum, or resultant, of two vectors such as two forces acting on a single object is the single vector that could replace the two and have the same effect.

There are three steps to finding the sum of two vectors. These are illustrated by working out the sum of a 2 N force acting up the page and a 3 N force acting from left to right, both forces acting on the same object.

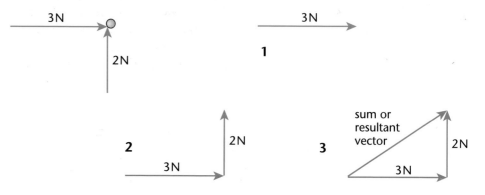

The order in which you draw the vectors does not matter, you should arrive at the same answer whichever one you start with.

- Draw an arrow that represents one of the vectors in both size and direction.
- Starting where this arrow finishes, draw an arrow that represents the second vector in size and direction.

- The sum, or resultant of the two vectors is represented (in both size and direction) by the single arrow drawn from the **start** of the first arrow to the **finish** of the second arrow.

In the example given, the size of the resultant force is 3.6 N and the direction is at an angle of 34° to the 3 N force. These figures were obtained by scale drawing.

Variations on a theme

AQA A	M2	NICCEA	M1
AQA B	M1	OCR A	M1
EDEXCEL A	M1	OCR B	M2
EDEXCEL B	M1	WJEC	M1

The method described above can be used to find the sum of any number of vectors by drawing an arrow for each vector, starting each new arrow where the previous one finished. The resultant of the vectors is then represented by the single arrow that starts at the beginning of the first vector and ends where the last one finishes. If the vectors being added together form a closed figure, i.e. the last one finishes where the first one starts, it follows that the sum is zero. This is what you would expect to find when working out the resultant force at a point in a stable structure for example.

For equilibrium at point X, the sum of the forces must be zero.

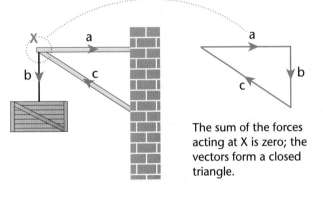

The sum of the forces acting at X is zero; the vectors form a closed triangle.

Although a scale drawing is often the quickest way of working out the resultant of two vectors, the size and direction can also be calculated. This is straightforward when the vectors act at right angles, but needs more complex mathematics in other cases.

In the worked example:

- the size of the resultant force can be calculated using Pythagoras' theorem as $\sqrt{(2^2 + 3^2)}$
- the angle between the resultant force and the 3 N force can be calculated using the definition of tangent as $\tan^{-1}(2 \div 3)$.

The notation 'tan⁻¹' means 'the angle whose tangent is'.

Bold letters are used here to represent both the size and direction of a vector. When normal type is used, it represents the size of the vector only.

The method used to add two vectors is also used to subtract one vector from another. For example, an aircraft has velocity **a** and changes its velocity by an amount **b** so that its new velocity is **c**. The relationship between these vectors is **a** + **b** = **c**. To work out the change in velocity **b**, the relationship can be written as **b** = **c** − **a**, the new velocity − the old velocity. Vector **b** is found by adding (−**a**) to **c**. Vector −**a** is the vector equal in size to **a** but opposite in direction. Vector subtraction is illustrated in the diagram below.

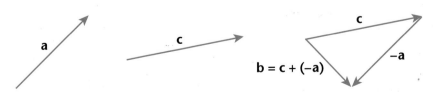

b = c + (−a)

Splitting a vector into two

A vector has an effect in any direction except the one at right angles to it. Sometimes a vector has two independent effects which need to be isolated. Just as the combined effect of two vectors acting on a single object can be calculated, two separate effects of a single vector can be found by splitting the vector into two **components**. Provided that the directions of the two components are chosen to be at right angles, each one has no effect in the direction of the other so they are considered to act independently.

The process of splitting a vector into two components is known as **resolving** or **resolution of** the vector.

The diagram shows the tension (T) in a cable holding a radio mast in place. The force is pulling the mast both vertically downwards and horizontally to the left.

> For a system in equilibrium the sum of all the vertical components is zero and the sum of all the horizontal components is zero.

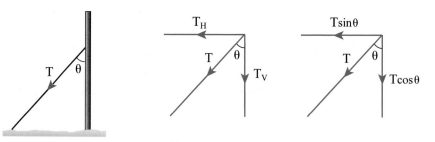

To find the effect of the tension in these directions, **T** can be split into two components, shown as T_H (the horizontal component) and T_V (the vertical component) in the diagram. You should check that according to the rules of vector addition, $T_H + T_V = T$.

> **KEY POINT**
>
> To find the magnitude of T_H and T_V the following rules apply:
>
> the component of a vector **a** at an angle θ to its own angle is $a\cos\theta$
>
> the component of a vector **a** at an angle $(90° - \theta)$ to its own angle is $a\sin\theta$

The application of these rules is shown in the right-hand diagram above.

Progress check

1 Work out the resultant of the two forces shown in the diagram.

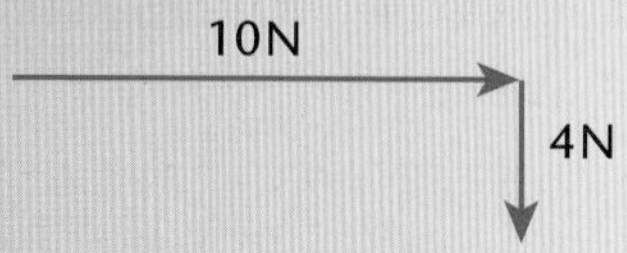

10N

4N

2 A car is being towed with a rope inclined at 20° to the horizontal. The tension in the rope is 350 N.
Work out the horizontal and vertical components of the tension in the rope.

3 A van is travelling north at a speed of 28 m s⁻¹.
After turning a corner it is heading 40° east of north at 25 m s⁻¹.
Work out the change in velocity of the van.

3 18 m s⁻¹ in a direction 29° S of E
2 Horizontal component = 329 N   Vertical component = 120 N
1 10.8 N at an angle of 22° to the 10 N force

1.2 Some effects of forces

After studying this section you should be able to:

- *calculate the moment of a force and apply the principle of moments to a stable object*
- *calculate the density of a solid, liquid or gas*
- *calculate the pressure due to a force*

LEARNING SUMMARY

The turning effect

AQA A	M2	NICCEA	M1
AQA B	M1	OCR A	M1
EDEXCEL A	M1	WJEC	M1

When forces are used to open doors, steer and pedal bicycles or turn a tap they are causing turning. The effect that a force has in turning an object round depends on:

- the size of the force
- the perpendicular (shortest) distance between the **force line** and the **pivot** (the axis of rotation).

Both of these factors are taken into account when measuring the turning effect, or **moment**, of a force.

The 'line of action' of a force is the line drawn along the direction in which the force acts.

> **Moment of a force = force × perpendicular distance from line of action of force to pivot.**
> The moment of a force, also known as torque, is measured in N m.
>
> **KEY POINT**

The diagram shows how the same force used to open a door can produce very different effects, according to the direction in which it is applied.

The moment of a force is a vector that can only have one of two directions; either clockwise or anticlockwise.

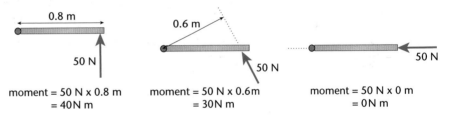

moment = 50 N x 0.8 m   moment = 50 N x 0.6m   moment = 50 N x 0 m
= 40N m   = 30N m   = 0N m

If the forces that make the couple are equal in size, the moment of the couple = size of one force × shortest (perpendicular) distance between the force lines.

When turning on a tap or steering a bicycle, two forces are normally used. The forces act in opposite directions, but they each produce a moment in the same direction. A pair of forces acting like this is called a **couple**. The combined moment is equal to the sum of the moments of the individual forces.

A question of balance

AQA A	M2	NICCEA	M1
AQA B	M1	OCR A	M1
EDEXCEL A	M1	WJEC	M1

For an object or a structure to be stable, two conditions apply:

- the resultant force in any direction is zero
- the resultant moment in either direction (clockwise or anticlockwise) is zero.

This second condition is known as the **principle of moments**.

> The principle of moments states that:
> For an object to be balanced, the sum of the clockwise moments about any pivot is equal to the sum of the anticlockwise moments about the same pivot.
>
> **KEY POINT**

The principle of moments can be used to determine the state of balance of an object and also to work out the size of the forces acting on objects that are known to be balanced.

Pressure

EDEXCEL A   M2
OCR A   M1
OCR B   M1

When a force is applied to a solid object such as a drawing pin or a nail, the force is transmitted through the solid. Factors that determine the effect of the force in cutting or piercing include:

- the size of the force
- the area of contact.

These two factors together are used to calculate the pressure.

If the force that causes the pressure is not acting in a direction normal to the surface, then the normal component of the force is used to calculate the pressure.

> **KEY POINT**
>
> Pressure = normal force per unit area
>
> $$p = F/A$$
>
> Pressure is measured in pascal (Pa) where 1 Pa = 1 N m⁻²

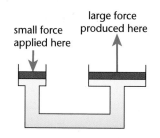

small force applied here | large force produced here

Whereas a solid transmits a force in the direction in which it is applied, fluids (liquids and gases) transmit pressure. The pressure at any point in a fluid has the same value in all directions. The transmission of pressure by a fluid is used in hydraulic machinery, where the force exerted by the fluid is determined by its pressure and the area over which it acts. This enables a large force to be exerted by the application of a small force that causes the pressure in the fluid. The principle of a hydraulic lifting device is shown in the diagram.

Density

AQA A   M3
EDEXCEL A   M1
EDEXCEL B   M2
OCR A   M1

The density of a material is a measure of how close-packed the particles are. Gases at atmospheric pressure are much less dense than solids and liquids because the particles are more widespread.

Typical densities of some everyday materials in kg m⁻³ are:

air 1.2
water 1000
concrete 5000
iron 9000

> **KEY POINT**
>
> Density of a material = mass per unit volume.
>
> $$\rho = m/V$$
>
> Density is measured in kg m⁻³ or g cm⁻³.

It is measured by dividing the mass of a specimen of the material by its volume.

Progress check

1 Use the principle of moments to calculate the tension in the rope, T, in the diagram below.

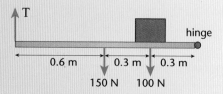

2 Explain how the hydraulic system shown in the diagram above multiplies the input force.

3 Describe how you would measure the density of air.

1.3 Under stress

After studying this section you should be able to:

- describe the difference between elastic and plastic behaviour
- state Hooke's law and appreciate its limitations
- contrast the behaviour of a ductile, a brittle and a polymeric solid when stretched
- explain how stress and strain are used in measurements of the Young modulus

LEARNING SUMMARY

Elastic or plastic

AQA A	M3	OCR A	M1
EDEXCEL A	M3B	OCR B	M1
EDEXCEL B	M2	WJEC	M1
NICCEA	A2		

No material is absolutely rigid. Even a concrete floor changes shape as you walk across it. The behaviour of a material subjected to a tensile (pulling) or compressive (pushing) force can be described as either **elastic** or **plastic**.

> a material is **elastic** if it returns to its original shape and size when the force is removed
>
> a material is **plastic** if it does not return to its original shape and size when the force is removed

Most materials are elastic for a certain range of forces, up to the **elastic limit**, beyond which they are plastic. Plasticine and playdough are plastic for all forces.

Hooke attempted to write a simple rule that describes the behaviour of all materials subjected to a tensile force.

> If the extension is proportional to the stretching force, then doubling the force causes the extension to double.

KEY POINT

Hooke's law states that:

the extension of a sample of material is proportional to the stretching force
$$e \propto F$$
This can be written as $F = ke$
where k represents the stiffness of the sample and has units of N m⁻¹

> A polymeric solid is one made up of long chain molecules.

Metals and springs 'obey' Hooke's law up to a certain limit, called the **limit of proportionality**. For small extensions, the extension is proportional to the stretching force. Rubber and other **polymeric** solids do not show this pattern of behaviour.

The graphs below contrast the behaviour of different materials subjected to an increasing stretching force.

> Because the limits of proportionality and elasticity lie close together on the force-extension curve, the term 'elastic limit' is often used to refer to both points.

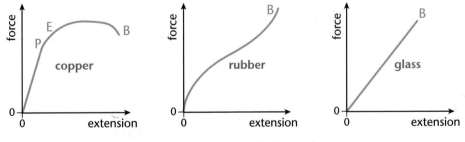

P = limit of proportionality   E = elastic limit   B = breaking point

- **Copper** is a **ductile** material, which means that it can be drawn into wires. It is also **malleable**, which means that it can be reshaped by hammering and bending without breaking. When stretched beyond the point E on the graph it retains its new shape.
- **Rubber** does not follow Hooke's law and it remains elastic until it breaks.
- **Glass** is **brittle**; it follows Hooke's law until it snaps.
- **Kevlar** is **tough**; it can withstand shock and impact.
- **Mild steel** is **durable**; it can withstand repeated loading and unloading.
- **Diamond** is **hard**; it cannot be easily scratched.

Storing energy

If seat belts and climbing ropes did not stretch, the forces exerted in stopping a person in an emergency would be too great for the body to withstand.

Some elastic materials are intended to absorb energy. The more a material is extended, the greater the force required.

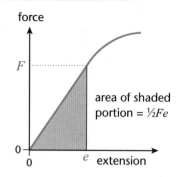

force

F

area of shaded portion = $\frac{1}{2}Fe$

e extension

- The energy stored as a material is deformed is represented by the area between the curve and the extension axis.
- If a force F produces an extension e below the limit of proportionality, then the energy stored = $\frac{1}{2}Fe$, as shown in the diagram opposite.
- since $F = ke$, energy stored = $\frac{1}{2}ke^2$.

Stress, strain and the Young modulus

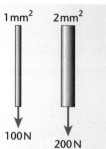

1mm²   2mm²

100N   200N

The stress on these wires is the same, even though the forces are different.

For some materials, e.g. copper and mild steel, the stress when the material breaks is less than the ultimate tensile stress.

The Young modulus for mild steel is 2.1×10^{11} Pa and that for copper is 0.7×10^{11} Pa. This means that for the same stretching force, a sample of copper stretches three times as much as one of mild steel with the same dimensions.

The extension that a force produces depends on the dimensions of the sample as well as the material that it is made from. To compare the elastic properties of materials in a way that does not depend on the sample tested, measurements of **stress** and **strain** are used instead of force and extension.

> **KEY POINT**
>
> stress = normal force per unit area
> $$\sigma = F/A$$
> Like pressure, stress is measured in Pa, where 1 Pa = 1 N m⁻²
>
> strain = extension per unit length
> $$\varepsilon = e/l$$
> Strain has no units.

The strength of a material is measured by its **breaking stress** or **ultimate tensile stress**. This is the maximum stress the material can withstand before it fractures or breaks and may not be the actual stress when breaking occurs. Its value is independent of the dimensions of the sample used for the test.

Using measurements of stress and strain, the stiffness of different materials can be compared by the **Young modulus**, E.

> **KEY POINT**
>
> Young modulus = stress ÷ strain
> $$E = \sigma/\varepsilon = F/A \div e/l = Fl/eA$$
> The Young modulus has the same units as stress, Pa.

The greater the value of the Young modulus the **stiffer** the material, i.e. the less it stretches for a given force.

Progress check

1. Describe the difference between a **ductile** material and one that is **brittle**.
2. A force of 150 N applied to a spring causes an extension of 25 cm.
   a. Calculate the energy stored in the spring.
   b. What assumption is necessary to answer **a**?
3. The Young modulus for copper is 1.3×10^{11} Pa.

   A tensile force of 150 N is applied to a sample of length 1.50 m and cross-sectional area 0.40 mm². (1.0 mm² = 1.0×10^{-6} m²).

   Assuming that the limit of proportionality is not exceeded, calculate the extension of the copper.

3   4.3 mm
b   the limit of proportionality has not been exceeded
2   a   18.75 J
1   A ductile material can be drawn into wires; a brittle material snaps easily.

1.4 Solid materials

After studying this section you should be able to:

- *describe the structure of metals and polymers and relate the structure to the properties*
- *explain how the properties of metals and polymers can be changed*
- *explain the advantages of using composite materials in appropriate applications*

LEARNING SUMMARY

Rubber

EDEXCEL A ▷ M3B   WJEC ▷ M1
EDEXCEL B ▷ M2

> The difference between a force–extension graph and a stress–strain graph for the same material is that they have different values and units on the axes.

Graphs that show the variation of stress with strain have the same shape as the force-extension graphs on page 31. The gradient of these graphs represents the stiffness of the material, i.e. how difficult the material is to stretch. The structure of **natural rubber** is like a can of worms, consisting of long chain molecules with occasional cross-linking between the chains. This is shown in the diagram below

The long chain molecules in unstretched and stretched rubber

When rubber is stretched the stiffness gradually decreases and it becomes relatively easy to stretch as the chains uncoil; once this has happened the rubber is much stiffer.

> If you repeatedly stretch and release a rubber band, you can feel the effect of heating caused by hysteresis.

The stress–strain graph for rubber (opposite) shows that the behaviour as a load is removed is not the same as that when the load is being increased. This is called **hysteresis** and the curves are said to make a **hysteresis loop**. In unit 1.3. Under stress, it was shown that the energy absorbed or released by a material is represented by the area between the curve and the extension axis of a force–extension graph. The equivalent area on a stress–strain graph represents energy per unit volume.

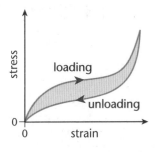

> **KEY POINT**
>
> On a graph of stress against strain:
> the area between the curve and the strain axis represents the energy per unit volume.

This is the energy absorbed when a material is being stretched and the energy that is released when the force is removed. Rubber absorbs more energy during loading than it releases in unloading. The difference is represented by the area of the hysteresis loop, shown shaded in the stress–strain graph.

> Most tyres use the same rubber to do two different jobs; provide rigidity for the tyre wall and elasticity for the tread.

The effect of hysteresis in rubber is to transfer energy to its molecules, resulting in heating. A car tyre goes through hysteresis for each revolution of the wheel. If the tyre is under-inflated this can lead to a 'blow-out' when the air inside the tyre becomes hot and the tyre explodes due to the increase in pressure. Modern, energy-efficient tyres reduce the problem of hysteresis by using a rigid rubber for the tyre walls and a more flexible rubber for the tread.

Steel

EDEXCEL A M3B   OCR B M1
EDEXCEL B M2

The behaviour of **steel** under stress depends on the carbon content of the metal. **High carbon steels** such as cast iron are brittle; they snap just beyond the elastic limit. **Mild steel** is tougher, much more energy is needed to cause failure. The diagram below compares stress–strain graphs for high carbon steel and mild steel.

> Mild steel can be deformed plastically, so it is used in place of wrought iron to make structures such as iron gates.

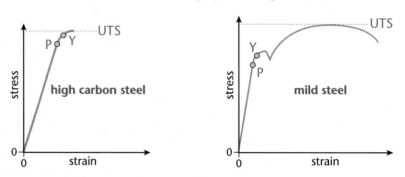

The points marked on the graphs are:

- P, the limit of proportionality; this is the end of the Hooke's law region
- Y, the elastic limit or yield point; beyond this point the material is permanently deformed
- UTS, the ultimate tensile stress or breaking stress; the maximum stress that the material can withstand.

> The term 'poly' means 'many' and 'crystalline' refers to atoms 'arranged in a regular array' to form crystals. The crystalline nature of metals can be seen clearly on the zinc coating on the inside of a can of pineapple.

Metals have a **polycrystalline** structure, which means that they consist of many small crystals. Within each crystal the atoms are arranged in a regular pattern. This structure is shown in diagram (a), which also shows the **slip planes** within the crystals. The ductile behaviour of mild steel beyond its elastic limit is due to whole planes of atoms slipping against each other. Slip occurs readily by the movement of an **edge dislocation**; a defect in the crystal due to an incomplete plane of atoms. This is shown in diagram (b).

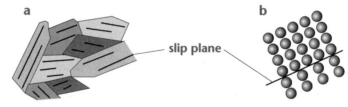

slip planes in metal crystals   the slip plane caused by a dislocation

> The 'neck' refers to the narrowing of the sample; this is sometimes called a 'waist'.

When slip occurs in a metal it can result in **necking**. If the width of the sample decreases at one point, the stress increases due to the reduction in cross-sectional area. The increased stress causes further stretching and narrowing at this point until the metal eventually fractures.

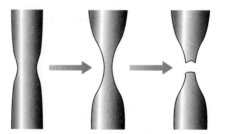

necking leading to fracture

Treating metals

EDEXCEL A M3B

Work hardening is not effective with metals which have low melting points such as lead (mp 328°C) and zinc (mp 420°C).

A ductile metal can be made stronger by preventing slip from occurring. In high carbon steel the presence of carbon atoms between layers of iron atoms stops them from slipping over each other. Strangely, the introduction of further dislocations can also strengthen metals such as copper or mild steel. This happens in a process called **work-hardening**, where the metal is repeatedly worked by hammering, stretching or bending while it is cold. The increased strength is attributed to a bottle-neck of dislocations preventing them from moving past each other.

When steel is heated to red heat, the grains enlarge. Slow cooling allows excess carbon to diffuse out of the crystals and results in the metal being annealed. Rapid cooling results in a brittle material as the excess carbon does not have time to diffuse out of the crystals.

The size of the crystals or grains in a metal also affects its strength. Since dislocations cannot move past the grain boundaries, smaller grains lead to stronger metals. Heat treatment of a metal affects both the size of the grains and the concentration of dislocations. It can be of several types:

- **annealing** involves heating a metal until it is red hot followed by controlled cooling. This makes the metal more ductile and less prone to fracture.
- **quenching** involves heating the metal until it is red hot and then cooling it rapidly by putting it in cold water or oil. This makes the metal strong but brittle.
- **tempering** is the process of quenching followed by annealing at a lower temperature. This is done to produce a balance between strength and toughness.

Failure

EDEXCEL A M3B

Lead pipes and roofing gradually become deformed under their own weight due to the effects of creep.

Creep can occur in metals and other materials subjected to a constant load, normally much lower than the yield stress, for a long time. Gradual deformation takes place due to a number of mechanisms, including:

- the movement of dislocations
- grain boundaries sliding relative to each other
- necking.

Creep affects structures such as the steel cables used to support suspension bridges, but it is more significant in metals used at temperatures that are above $0.3 \times$ the melting point of the metal in K. This is the case for lead used at room temperatures and also for metals used in engines. Creep-resistant alloys have been developed for use in applications where failure due to creep could be disastrous, for example, in the turbine blades used in aircraft jet engines.

Fatigue is another type of failure caused by a cumulative effect over time. It affects metals subjected to repeated cycles of stress such as that caused by stretching, rotation or vibration. Fatigue failure starts with a small crack appearing due to repeated working. Once a crack has formed, stress builds up around it due to the reduction in cross-sectional area. This causes the crack to enlarge until the material fractures.

The stress concentration at the tip of a crack increases as the crack grows.

The risks of fatigue failure can be reduced during both design and manufacture. In designing structures that are subjected to repetitive stress, sharp corners should be avoided as this is where cracks can develop. In the manufacturing process, care should be taken to avoid poor workmanship at places where materials are rivetted or bolted together leaving tiny cracks that then propagate.

Polymers

EDEXCEL A ▸ M3B   WJEC ▸ M1
EDEXCEL B ▸ M2

Amorphous describes any structure that is irregular, with no ordering. Glass has an amorphous structure.

Polymers consist of long chain molecules; natural rubber is a polymer whose molecules have no ordering. Such polymers are called **amorphous**. Perspex is a synthetic polymer whose structure is also amorphous. Nylon and polythene have both crystalline and amorphous regions, their structure is described as **semi-crystalline**. The diagram below shows how the chains in a semi-crystalline polymer are folded; the straight rows represent crystalline regions and the bends are amorphous.

Both nylon and polythene soften on heating because the bonds between the long chain molecules are weak; they are called **thermoplastics**. Because of this property they can be used to manufacture articles by moulding.

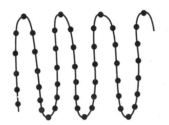

refolded chains in polythene

cross-linking makes rubber less stretchy

The rubber used in the manufacture of car tyres has a low sulphur content, about 5%. Hard rubbers contain up to 50% sulphur.

The rigidity of a polymer is increased by **cross-linking**, the formation of strong bonds between the molecules by introducing other substances. The vulcanised rubber used in car tyres contains sulphur which forms bonds between chains, making it less elastic. Adding more sulphur makes the rubber more rigid. As more cross-linking is employed, the polymer becomes a **thermoset**, one which hardens with increasing temperature.

Bakelite and melamine are thermosets commonly used for household objects.

Thermosetting polymers are used in the manufacture of electrical fittings such as lamp holders, sockets and kitchen worktops. Because they cannot be softened by heating, any moulding has to be done as the polymer is manufactured.

Composite materials

EDEXCEL A ▸ M3B

A **composite** is a mixture of two or more materials bonded together to obtain some overall desirable property.
Types of composite include:

* laminated materials
* fibre composites
* particle composites.

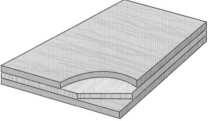

the grain in layers of plywood

Wood and bone are natural composite materials. Wood is a fibre composite, consisting of fibres of cellulose in a lignin cement.

A **laminate** consists of thin sheets of material joined together. Common examples are **plywood** and car windscreens. Wood is strong when subjected to a tensile force parallel to its grain, but if the force is across the grain it splits easily. The diagram above shows how plywood is manufactured by glueing sheets together so that the grains in adjacent sheets are at right angles to each other.

A windscreen is made in a similar way; two sheets of toughened glass are bonded to a sheet of clear plastic in a **sandwich**. If the glass shatters, the plastic holds the broken pieces together.

Chipboard is a **fibre composite** that consists of wooden chips glued together. The packing of the chips is random, so the wooden fibres lie in all directions. Unlike natural wood, chipboard does not have a grain so it is equally strong in all directions.

Fibre composites such as glass reinforced plastic (glass fibre) and carbon fibre reinforced plastics have replaced metals in some applications. They are lighter than metals and are not affected by fatigue. Some car bodies are made from glass fibre. Its low weight gives improved acceleration and fuel consumption but it is not as strong as the same thickness of steel, so it gives less protection to the driver and passengers.

The diagram below shows two ways in which glass fibres can be arranged within the plastic. When the fibres are parallel the glass fibre is strong in the direction of the fibres but weak in the direction at right angles to them. A random arrangement of the fibres gives a material that is equally strong in all directions.

> Glass fibre does not corrode like metals, so it is also suitable for applications such as the hulls of boats. Because glass fibre is non-magnetic, it is particularly suitable for minesweepers.

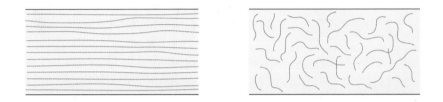

Two ways of embedding glass fibres in plastic

> In normal use glass breaks easily. This is because its surface is easily scratched and these scratches propagate. By protecting the surface from scratches, the ultimate tensile strength of glass is increased.

Carbon fibre reinforced plastic has the same tensile strength as steel but is much stiffer. It is still a very expensive material to manufacture so its use is limited to some parts of jet engines and such things as specialist tennis rackets.

Both types of fibre reinforced plastic rely on the intrinsic strength of carbon and glass. These materials are very brittle and become much weaker when the surface is scratched. The plastic protects the surface and prevents any cracks that do form from propagating along other fibres.

Using concrete

EDEXCEL A M3B   WJEC   M1

> The term aggregate applies to any filler material used in concrete. Sand is the finest aggregate used; coarse aggregates include gravel and crushed stone.

Concrete is a **particle composite** material made from a mixture of cement, water and aggregates. The cement bonds together the sand and small stones that make up the aggregate. This makes a cheaper material than using cement on its own.

sand and cement mixture

stones

The structure of concrete

Concrete is very strong when subjected to a compressive force, making it a suitable material for the pillars that support a road bridge. It breaks easily when in tension, so it is not suitable for the horizontal surfaces supported by the pillars, parts of which are in compression and parts of which are in tension.

To overcome this, concrete can be pre-stressed which ensures that the concrete remains under compression even when subjected to its normal tensile load. In the manufacture of pre-stressed concrete, the concrete is cast around steel bars that are held in tension.

Concrete that is always in compression in normal use is not pre-stressed, but it can be reinforced by casting the concrete around a steel mesh.

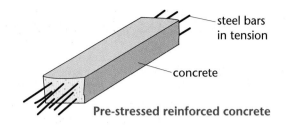

steel bars in tension

concrete

Pre-stressed reinforced concrete

After the concrete has hardened the tension is released, shrinking the steel and compressing the concrete. The steel both **reinforces** the concrete, sharing the load, and keeps it in compression even when it is stretched.

Forces in structures

EDEXCEL A   3B   OCR A   M1

The resultant force and resultant moment both have to be zero for the structure to be stable.

Resolving horizontal and vertical components will always give the same answers as taking moments.

If a beam or rod is **uniform** its mass is evenly distributed so that it is the same as if all the mass was concentrated at a single point at the centre of the beam. The weight is the force acting vertically downwards from this point.

The resultant force and resultant moment at any point in a stable structure are both zero. Either of these principles can be used to find the values of forces F_1 and F_2 in the ropes that support the beam shown in the diagram below.

- Both F_1 and F_2 could be found by resolving the vertical components of all the forces (which must balance) and the horizontal components of all the forces (which must also balance).
- Since F_1 does not have a moment about A, taking moments about this point gives a relationship between F_2 and the forces we know.
- Similarly, taking moments about B gives a relationship between F_1 and the known forces.

To take moments about A, F_2 needs to be resolved into horizontal and vertical components. The moment of the horizontal component is zero and that of the vertical component is in the anticlockwise direction. The equation is:

$$F_2 \cos 60° \times 4.0 \text{ m} = 200 \text{ N} \times 2.0 \text{ m} + 80 \text{ N} \times 1.0 \text{ m}$$
$$F_2 = 480 \text{ N m} \div (\cos 60° \times 4.0 \text{ m}) = 240 \text{ N}.$$

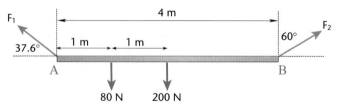

Progress check

1 Natural rubber is an **amorphous polymer**.
   a Explain the meaning of the terms **amorphous** and **polymer**.
   b Explain why rubber becomes stiffer as it is stretched.

2 What effect do each of the following have on a metal?
   a Work-hardening.
   b Annealing.

3 Use the principle of moments to calculate the value of force F_1 in the diagram above.

1.5 Motion in one and two dimensions

After studying this section you should be able to:

The term 'curve' means the line drawn on the graph to show the relationship between the quantities. It could be straight or curved.

- interpret graphs used to represent motion, and understand the physical significance of the gradient and the area between the curve and the time axis
- apply the equations that describe motion with uniform acceleration
- analyse the motion of a projectile

LEARNING SUMMARY

Speed and velocity

AQA A	M2	NICCEA	M1
AQA B	M1	OCR A	M1
EDEXCEL A	M1	OCR B	M2
EDEXCEL B	M1	WJEC	M1

Information about the **speed** of an object only states how fast it is moving; the **velocity** also gives the direction. For an object moving along a straight line, positive (+) and negative (−) are usually used to indicate movement in opposite directions. As long as it is moving, the **distance** travelled by the object is increasing, but its **displacement** can increase or decrease and, like velocity, can have both positive and negative values.

The fixed point is usually the starting point of the motion or the object's rest position.

- **displacement** is a vector quantity; it specifies both the distance and direction of an object measured from a fixed point.

The diagram shows a distance–time graph and a displacement–time graph for the same motion.

Check that you understand why distance has only positive values, but displacement has both positive and negative values.

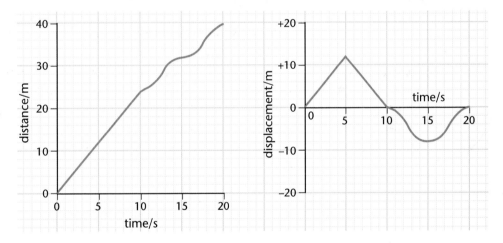

The same symbols are used for speed and velocity. When an object is changing speed/velocity, u is often used for the initial speed/velocity, with v being used for the final speed/velocity. The symbol c is usually reserved for the speed of light.

Information from the graphs can be used to calculate the **average speed** and **average velocity** over any time period, using the relationships:

average speed	= distance travelled	÷ time taken	$v = \Delta d \div \Delta t$
average velocity	= displacement	÷ time taken	$v = \Delta s \div \Delta t$

KEY POINT

These show that:

- over the 20 s time period, the average speed is 2 m s⁻¹ and the average velocity is 0
- the average speed is constant for the first 10 s, after which it changes.

Calculations of the average speed or velocity over a period of time do not show any changes that have taken place. **Instantaneous** values are defined in terms of **the rate of change** of distance or displacement and are represented by the gradients of the graphs.

> instantaneous speed = rate of change of distance. It is represented by the gradient of a distance–time graph.
>
> instantaneous velocity = rate of change of displacement. It is represented by the gradient of a displacement–time graph.

KEY POINT

The gradient of a displacement–time graph can have either a positive or a negative value, so it shows the direction of motion as well as the speed.

Speed–time and velocity–time graphs

AQA A	M2	NICCEA	M1
AQA B	M1	OCR A	M1
EDEXCEL A	M1	OCR B	M2
EDEXCEL B	M1	WJEC	M1

Any object that is changing its speed or direction is **accelerating**. Speeding up, slowing down and going round a corner at constant speed are all examples of **acceleration**. Acceleration is a measure of how quickly the velocity of an object changes.

> In physics, any change in velocity is an *acceleration*. This means that an object moving at constant speed, but changing direction, is accelerating.

> Instantaneous acceleration = rate of change of velocity. It is represented by the gradient of a velocity–time graph.
>
> average acceleration = change in velocity ÷ time $\qquad a = \Delta v \div \Delta t$
>
> Acceleration is a vector quantity and is measured in m s^{-2}.

KEY POINT

Speed–time and velocity–time graphs both give information about the motion of an object that is accelerating. The graphs below both represent the motion of a ball after it has been thrown vertically upwards with an initial speed of 15 m s^{-1}.

> The 'negative' and 'positive' velocities show motion in opposite directions; in this case 'positive' is up and 'negative' is down.

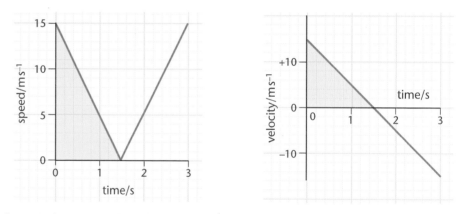

The speed–time graph shows that the speed decreases to zero as the ball reaches its maximum height and then increases. The velocity–time graph also shows the change in direction of the ball. In each case the gradient of the graph is numerically equal to the acceleration (as the ball is moving vertically this is free-fall acceleration), but the gradient of the velocity–time graph also shows that the direction of the acceleration (vertically downwards) is opposite to that of the initial velocity.

> Remember, the area of a triangle = ½ × base × height.

The distance that the ball travels in any time interval can also be obtained from the graphs. Since **distance travelled = average speed × time**, this is represented by the area between the curve and the time axis. The area that represents the distance travelled by the ball while moving upwards is shaded on each graph.

> The gradient of a velocity–time graph represents the acceleration.
>
> The area between the curve and the time axis of a velocity–time or a speed–time graph represents the distance travelled.

KEY POINT

Check that the shaded area represents a distance of 11.25 m in each case.

Uniform acceleration

The gradient of the velocity–time graph on page 40 has a constant value; it represents a constant or **uniform** acceleration. A number of equations link the variable quantities when an object is moving with uniform acceleration. They are:

- $v = u + at$
- $s = ut + \frac{1}{2}at^2$
- $v^2 = u^2 + 2as$
- $s = \frac{1}{2}(u + v)t$

> The symbols have the meanings already defined:
>
> u = initial velocity
> v = final velocity
> s = displacement
> a = acceleration
> t = time

These equations involve a total of five variables but only four appear in each one, so if the values of three are known the other two can be calculated.

> To work out the distance travelled by an object that changes direction, calculate the displacement both before and after the change in direction, and add these together.

When using the equations of uniformly accelerated motion:

- take care with signs; use + and – for vector quantities such as velocity and acceleration that are in opposite directions
- remember that s represents displacement; this is not the same as the distance travelled if the object has changed direction during the motion.

Motion in two dimensions

People who play sports such as tennis and squash know that, no matter how hard they hit the ball, they cannot make it follow a horizontal path. The motion of a ball through the air is always affected by the Earth's pull.

> There is a simple relationship between the total distances travelled after t, $2t$ etc. by an object falling vertically. Can you spot it?

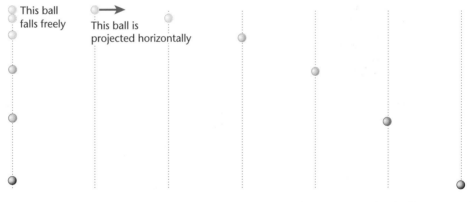

This ball falls freely

This ball is projected horizontally

The diagram shows the results of photographing the motion of a ball projected horizontally alongside one released so that it falls freely. The photographs are taken at equal time intervals.

The free-falling ball travels increasing distances in successive time intervals. This is because the vertical motion is accelerated motion, so the average speed of the ball over each successive time interval increases.

The ball projected horizontally travels equal distances horizontally in successive time intervals, showing that its horizontal motion is at **constant speed**. Vertically, its motion matches that of the free-falling ball, showing that **its vertical motion is not affected by its horizontal motion**.

> The motion of any object can be resolved in two directions at right angles to each other. The two motions can then be treated separately.

> **KEY POINT**
> When an object has both horizontal and vertical motion, these are independent of each other.

This important result means that the horizontal and vertical motions can be analysed separately:

- for the horizontal motion at constant speed, the equation $v = s \div t$ applies
- for the vertical, accelerated, motion, the equations of motion with uniform acceleration apply.

Solving projectile problems

The common factor between the two independent motions of an object that is travelling both vertically and horizontally is t; the time for which it is travelling. It is important not to confuse the horizontal velocity and displacement with those that apply to the vertical motion.

Consider this problem.

A ball is thrown horizontally with a speed of 15.0 m s^{-1} from a vertical height of 18.0 m. How far has it travelled horizontally when it reaches the ground? Take $g = 10.0$ m s^{-2}.

To solve this problem:

> A common error in solving this type of problem is to take u, the initial vertical velocity, to be equal to the value of the horizontal velocity.

- apply the equations of motion with uniform acceleration to the vertical motion to find the time that the ball is in the air. Note that vertically $u = 0$.
- use the constant speed equation to calculate the distance travelled horizontally by the ball in this time.

Check that this gives an answer of 27 m.

Progress check

1 The diagram below shows a velocity–time graph. Calculate the distance travelled and the acceleration on each labelled section of the graph.

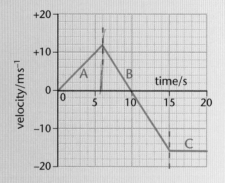

> Hint for Q2: what is the velocity of the pigeon at its maximum height?

2 A clay pigeon is fired upwards with an initial velocity of 23 m s^{-1}. What height does the pigeon reach? Take $g = 10$ m s^{-2}.

3 A tennis player is standing 5.5 m from the net. She hits the ball horizontally at a speed of 32 m s^{-1}. How far has the ball dropped when it reaches the net? Take $g = 10$ m s^{-2}.

3 0.15 m
2 26.45 m
   c 80 m; 0 m s^{-2}
   b 64 m; -3.1 m s^{-2}
1 a 36 m; 2 m s^{-2};

Sample question and model answer

A wooden pole is held in an upright position by two wires in tension. These forces are shown in the diagram below.

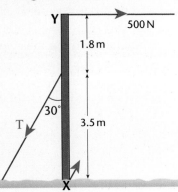

The pole is fixed at its base.

(a) Calculate the moment of the 500 N horizontal force about X. [2]

Always write down the formula that you intend to use, and each step in the working. It would be very easy in this case to use a wrong distance. Simply writing down a wrong answer always gets no marks, but showing how you arrive at that wrong answer allows credit to be awarded for the correct process.

moment = force x perpendicular distance to pivot
= 500 N x 5.3 m   1 mark
= 2650 N m   1 mark

(b) Calculate the value of the force T. [3]

Only the horizontal component of *T* has a turning effect; its moment must counterbalance that of the 500 N horizontal force.

The horizontal component of the force T must also have a moment of 2650 N m: 1 mark

Having written down the physical principle, the next step is to write it in the form of an equation. *T*cos60° or *T*sin30° is the horizontal component of *T*.

Tcos60° x 3.5 m = 2650 N m   1 mark

The final step is to calculate the value of *T*.
Although it is not essential to show the units of physical quantities at each step in the calculation, it is good practice and you MUST include the correct unit with your answer to each part.

T = 2650 N m ÷ (cos60° x 3.5 m) = 1514 N   1 mark

This problem cannot be solved by resolving *T* and the 500 N force as there is a force at X (you can see that this is not just a vertical reaction force by taking moments about Y – there must be a horizontal component with a moment that balances the horizontal component of *T*).

Practice examination questions

Take the value of free-fall acceleration, *g*, to be 10 m s⁻².

1 A microlight aircraft heads due north at a speed through the air of 18 m s⁻¹.
A wind from the west causes the aircraft to move east at 6 m s⁻¹.
Draw a vector diagram and use it to work out the resultant velocity of the
aircraft.   [3]

2 The diagram shows a light fitting held in place by two cables.

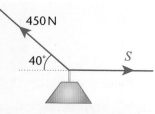

(a) Calculate the vertical component of the 450 N force.   [2]

(b) Write down the weight of the light fitting.   [1]

(c) Explain why a vector diagram that represents the forces on the light fitting
is a closed triangle.   [2]

(d) Draw the vector diagram and use it to find the value of the force *S*.   [2]

3 The diagram shows a garden tool being used to cut through a branch.

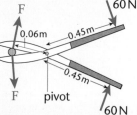

(a) Calculate the moment of the force on each handle.   [2]

(b) Assuming equilibrium, calculate the value of the resistive force, *F*, that
acts on each blade.   [2]

(c) Suggest why a similar tool designed to cut through thicker branches has
longer handles.   [2]

Practice examination questions (continued)

4 The diagram shows the base of a steel girder that supports the roof of a building. The downward force in the girder is 4500 N and it is fixed to a square steel plate, each side of which is 0.30 m long.

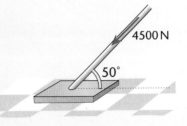

(a) Calculate the values of the horizontal and vertical components of the force in the girder. [2]

(b) Calculate the pressure that the steel plate exerts on the ground. [3]

(c) Explain why the girder is fixed to a steel plate rather than being fixed directly to the ground. [2]

5 A large aquarium has a horizontal cross-section of 10.5 m × 4.2 m. It is filled with water to a depth of 3.5 m. The density of water is 1000 kg m⁻³.

(a) Calculate the weight of the water in the aquarium. [3]

(b) Calculate the pressure on the base of the aquarium due to the weight of the water. [3]

(c) Explain why the glass used for the sides of the aquarium needs to be as strong as the base. [2]

(d) A similar aquarium has a horizontal cross-section of 10.5 m × 2.1 m and is filled with water to the same depth.
How does the pressure on the base compare to that calculated in b?
Give the reason for your answer. [2]

6 (a) How does a ductile material behave differently from a brittle one when subjected to a stretching force? [2]

(b) The graph shows the relationship between the stress on a sample of cast iron and the strain that it causes. The curve ends when the sample breaks.

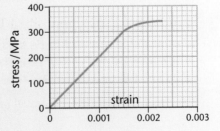

(i) Write down the value of the breaking stress of the cast iron. [1]

(ii) For what range of stresses does the cast iron follow Hooke's law? [1]

(iii) Calculate the Young modulus for the cast iron in the region where it follows Hooke's law. [2]

(iv) A second sample of the same material has twice the diameter of the original sample.
Explain whether the stress–strain graph for this sample would be the same as that shown in the diagram. [2]

Practice examination questions *(continued)*

7 The diagram is a velocity–time graph for a train travelling between two stations. A positive velocity represents motion in the forwards direction.

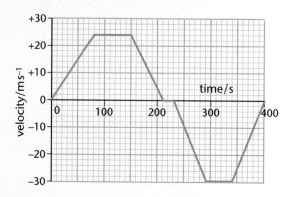

(a) Calculate the acceleration of the train during the first 80 s. [3]

(b) During which other times shown on the graph is the train increasing speed? [1]

(c) For what length of the time was the train:
   (i) not moving? [1]
   (ii) travelling at a constant non-zero speed? [1]

(d) Calculate the total distance travelled by the train. [2]

(e) At the end of the time shown on the graph, how far was the train from its starting point? [1]

8 In a game of cricket, a ball leaves a bat in a horizontal direction from a height of 0.39 m above the ground.
The speed of the ball is 35 m s^{-1}.
(a) Calculate the time interval between the ball leaving the bat and reaching the ground. [3]

(b) What horizontal distance does the ball travel in this time? [2]

(c) Fielders are not allowed within 15 m of the bat.
What is the maximum speed that the ball should leave the bat, travelling in a horizontal direction, for the player not to be caught out? [2]

9 A motorist travelling at the legal speed limit of 28 m s^{-1} (60 mph) takes his foot off the accelerator as he passes a sign showing that the speed limit is reduced to 14 m s^{-1} (30 mph). The car decelerates at 2.0 m s^{-2}.
(a) For what time interval is the motorist exceeding the speed limit? [2]

(b) How far does the car travel in that time? [2]

10 A fountain is designed so that the water leaves the nozzle and rises vertically to a height of 3.5 m.
(a) Calculate the speed of the water as it leaves the nozzle. [3]

(b) For how long is each drop of water in the air? [2]

1.6 Force and acceleration

After studying this section you should be able to:

- recall and use the relationship between resultant force, mass and acceleration
- explain the principles of seat belts, air bags and crumple zones
- interpret and explain data relating to vehicle stopping distances

Getting going

AQA A	M2	NICCEA	M1
AQA B	M1	OCR A	M1
EDEXCEL A	M1	OCR B	M2
EDEXCEL B	M1	WJEC	M1

The force that pushes a cycle or other vehicle forwards is called the driving force, or motive force.

Changing motion requires a force. Starting, stopping, getting faster or slower and changing direction all involve forces. One way to start an object moving is to release it and allow the Earth to pull it down. The Earth's pull causes the object to accelerate downwards at the rate of 10 m s^{-2}.

To set off on a cycle, you push down on the pedal. The chain then transmits this force to the rear wheel. The wheel pushes on the road and, provided that there is enough friction to prevent the wheel from slipping, the push of the road on the wheel accelerates the cycle forwards.

However, as every cyclist knows, this acceleration is not maintained. The cyclist eventually reaches a speed at which the driving force no longer causes the cycle to accelerate. This is due to the **resistive** forces acting on the cycle. The main one is **air resistance**, although other resistive forces act on the bearings and the tyres.

Air resistance or **drag** also affects the motion of an object released and allowed to fall vertically; in fact it opposes the motion of anything that moves through the air. This is due to the air having to be pushed out of the way. The faster the object moves, the greater the volume of air that has to be displaced each second and so the greater the resistive force.

The diagram shows the directions of the forces acting on a cyclist travelling horizontally and a ball falling vertically.

The single arrow drawn acting on the front of the cyclist represents all the resistive forces acting against the motion.

resistive forces

driving force

air resistance

weight

The effect of the resistive forces is to reduce the size of the resultant, or unbalanced, force that causes the acceleration. How this affects the speed of the cycle and ball is shown by the graph.

The decreasing gradient of the speed–time graph shows that the acceleration decreases as the resultant force becomes smaller. At a speed where the resistive force is equal to the driving force, the resultant force is zero and the object travels at a constant speed.

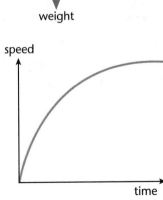

speed

time

When the forces acting on a moving object are balanced, the constant speed is called its terminal speed or terminal velocity.

To summarise:
- an object accelerates if there is a resultant, or unbalanced, force acting on it
- the acceleration is proportional to the resultant force and in the direction of the resultant force.

What else affects the acceleration?

> Galileo's insistence that heavy objects do not fall faster than lighter ones cost him his job as professor of mathematics at the University of Pisa.

Galileo appreciated that heavy objects have the same free-fall acceleration as lighter ones, provided that they are not affected by air resistance. You can confirm this by releasing two different-sized coins side-by-side; at low speeds, air resistance has little effect so you see and hear them reaching the ground together.

In vertical motion the **mass** of an object affects the force causing the acceleration; the heavier the object, the greater the force pulling it down. Doubling the mass doubles the pulling force (since $W = m \times g$) **but does not affect the acceleration.**

> If the mass of an object is doubled, its acceleration is halved for the same pulling force.

This is not the case in horizontal motion, where increasing the mass of an object has no effect on the force causing it to accelerate, but does affect the acceleration. As the number of passengers on a bus increases, its acceleration away from the bus stop decreases.

Results of experiments using a constant force to accelerate different masses show that:

> the acceleration of an object is inversely proportional to its mass

> The symbol Σ, meaning 'sum of' is used here to emphasise that the relationship applies to the resultant force on an object, and not to individual forces.

The dependence of acceleration on both the resultant force and the mass is summarised by the relationship:

> **KEY POINT**
>
> resultant force = mass × acceleration
> $$\Sigma F = ma$$
> Where the unit of force, the newton, is defined as the force required to cause a mass of 1 kg to accelerate at 1 m s⁻².

Forces and braking

Vehicles that travel on wheels rely on friction between the tyres and the road surface to provide the driving force. Ice on a road surface and wet leaves on railway lines can reduce the friction force, resulting in the wheels spinning round as they slide over the surface.

Friction is also used in braking systems to bring cycles and motor vehicles to a halt. Two common types of brake used are **drum brakes** and **disc brakes**. These are shown in the diagram.

> Most modern vehicles use disc brakes because of their greater efficiency.

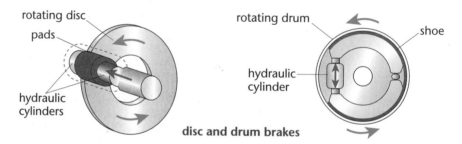

disc and drum brakes

When the driver applies the brakes, high friction material in the shoe or pad is pressed against the drum or disc. The friction force exerts a torque in the opposite direction to the torque that drives the wheel. During braking the tyre pushes **forwards** on the road surface, causing the road surface to push **backwards** on the tyre. The resultant force in the backwards direction causes the vehicle to decelerate.

> A deceleration is a negative acceleration. In this case the acceleration is in the opposite direction to the vehicle's velocity, so it has the opposite sign.

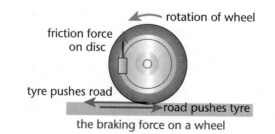

the braking force on a wheel

Braking forces, like driving forces, rely on friction. The size of the friction force depends on the nature of both the road and the tyre surface. The greater the friction force, the shorter the distance required to stop the vehicle.

Grand prix drivers match their tyres to the road conditions. Excessive friction wastes fuel and reduces the acceleration and top speed. However, having fitted tyres suitable for dry road conditions, a sudden shower of rain can suddenly send the cars out of control!

Tyre **tread** is designed to prevent a layer of water building up between the tyre and the road. This reduces the friction force, reducing the effectiveness of steering, drive and braking. If there is insufficient tread on a tyre it cannot push the water out of the way quickly enough, as the channels in the tyres are too small for all the water to pass through.

Car safety

OCR A   M1

If there is not enough friction, a passenger appears to be 'thrown' in the opposite direction to the change in velocity.

When the velocity of a car changes, that of the people inside it has to change too. During normal driving, the force to accelerate the driver and passengers comes from the seat; it pushes the person forwards to cause an increase in speed and the friction force between the person and the seat is sufficient to slow the person down or cause a change in direction.

During hard braking or in a collision, friction is not enough to match the car's deceleration. Viewed from the outside of the car, a passenger appears to have been thrown forwards. Despite what the television advertisements say, this is **not** the case! The photograph shows what can happen to a passenger in a collision. **Seat belts** prevent this from happening. However, a rigid seat belt that caused a passenger to decelerate at the same rate as the car could prove fatal in a head-on collision.

Excessive force from a seat belt can break the ribs and damage internal organs.

To be effective a seat belt should:
- restrain the passenger and prevent collisions with the inside of the car
- allow the passenger to come to rest over a longer time period than that taken by the car.

To achieve these conflicting aims, seat belts are designed so that they stretch sufficiently to allow the passenger to carry on moving for a short time after the car has stopped, but not so much that would result in the passenger hitting the windscreen or the seat in front.

There is less space in front of the driver than there is in front of the passengers because of the steering wheel. In modern cars this is designed to collapse on impact.

Seat belts can still inflict severe injuries during a collision. The amount of space to allow stretching, particularly in front of a driver, is limited so the restraining force from the belt can be large. There is also a design conflict in deciding on the width of the belt; wide belts exert less pressure than narrow ones but are less comfortable to wear.

Airbags are mis-named as they do not use air. An inert gas such as nitrogen or argon is used to inflate the bag within 0.02 s of a collision.

Airbags are designed to restrain a driver and passengers without any risk of causing physical damage to the person. The bag only operates during a rapid deceleration such as a head-on collision. This releases a gas which causes the bag to inflate and surround the driver or passenger like a cushion. The force exerted on the person is similar to that from a seat belt, but because it is over a much larger area the pressure is smaller.

Modern cars, and railway carriages, are also designed so that parts of them crush on impact. Although the part of the vehicle containing passengers is rigid to give them protection, the front and rear are crumple zones.

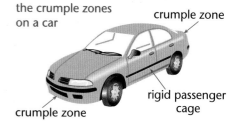

the crumple zones on a car

crumple zone

crumple zone

rigid passenger cage

As the name implies, the crumple zone should squash during a collision. This has two effects:

- it increases the time it takes to stop the vehicle, reducing the deceleration and the force
- it absorbs much of the kinetic energy of the moving vehicle, so that it does not bounce backwards, causing further injury to passengers.

Forces and towing

OCR A   M1

The term 'acceleration' is used here with the everyday meaning, i.e. speeding up.

Towing a caravan or trailer affects the steering, acceleration and braking of a vehicle.

When accelerating, there is more mass to be accelerated, so the maximum acceleration is reduced. The force that accelerates the towed vehicle is transmitted through the tow bar, which is in tension. As well as pulling forwards on the towed vehicle, the tow bar also pulls backwards on the car, reducing the resultant force that causes acceleration.

Tension in the tow bar is maintained when the car and towed vehicle are travelling at a steady speed, but now the force is reduced, being equal to the resistive force acting on the towed vehicle.

During braking, the car pushes back on the vehicle being towed, which in turn pushes forwards on the car. This reduces the car's deceleration, so braking distances are increased. The diagram shows the forces in the tow bar when the car and vehicle being towed are accelerating and braking.

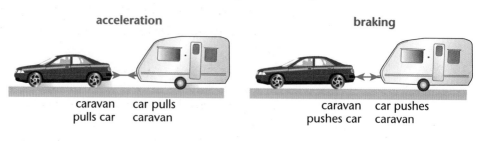

acceleration

caravan pulls car

car pulls caravan

braking

caravan pushes car

car pushes caravan

Stopping distances

OCR A   M1

In the highway code there is a chart that shows the stopping distances in good conditions for a car travelling at different speeds. There are two components to the stopping distance at a particular speed, thinking distance and braking distance.

Thinking distance is how far the vehicle travels while the driver is reacting. The two factors that determine this are:
- reaction time
- vehicle speed.

Drugs, including alcohol, nagging children and loud hi-fi can all increase a driver's reaction time.

A driver's reaction time depends on a number of things, including the state of the driver and the nature of the hazard, but a driver should be able to react to an unexpected hazard within 0.6 s.

As the diagram shows this means that doubling the vehicle speed doubles the thinking distance.

During the driver's reaction time the vehicle continues at a constant speed, so **distance travelled = speed × time**. Assuming a constant reaction time, this means that:

thinking distance is proportional to speed.

This is not the case with braking distance. As with reaction time, a number of factors could affect this, including the nature and condition of the road surface, the tyres and the brakes. The following calculations to work out the braking distances from speeds of 10 m s^{-1} and 20 m s^{-1} assume a deceleration of 8.3 m s^{-2}.

initial speed/m s^{-1}	10	20
time taken to brake/s	1.2	2.4
average speed during braking/m s^{-1}	5	10
braking distance/m = average speed × time	6	24

This shows that doubling the speed quadruples the braking distance. Not only is the average speed during braking doubled, so is the time it takes to stop, so:

braking distance is proportional to (speed)^2.

The diagram shows how stopping distance is related to speed:

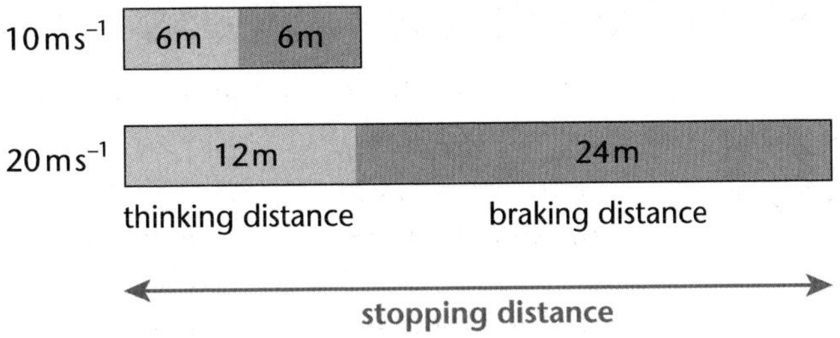

The overall effect of speed on stopping distance is approximately 'doubling the speed trebles the stopping distance'.

10 m s^{-1} | 6m | 6m

20 m s^{-1} | 12m | 24m

thinking distance   braking distance

stopping distance

Progress check

1 a The total mass of a cyclist and her cycle is 85 kg.
   Calculate the acceleration as she sets off from rest if the driving force is 130 N.
  b The driving force remains constant at 130 N. What is the size of the resistive force when she is travelling at terminal velocity?

2 A fully-laden aircraft weighs 300 000 kg. It accelerates from rest to a take-off speed of 70 m s^{-1} in 20 s.

  a Calculate the acceleration of the aircraft.
  b Calculate the size of the resultant force needed to cause this acceleration.
  c Explain why the force from the engines must be greater than the answer to **b**.

3 A car is travelling at the legal maximum speed of 28 m s^{-1} on a single carriageway. The driver notices a hazard in the road ahead.

  a If the driver's reaction time is 0.5 s and the car decelerates at 7 m s^{-2} during braking, calculate the stopping distance.
  b Calculate the size of the force required to decelerate the car if its total mass is 850 kg.

1 a 1.53 m s^{-2}
  b 130 N
2 a 3.5 m s^{-2}
  b 1.05 × 10^6 N
  c the force also has to act against air resistance and friction.
3 a 70 m
  b 5950 N

1.7 Newton I and III

After studying this section you should be able to:

- identify the forces acting on an object that is not changing its velocity
- for a given force, describe the other force that makes up the third law 'pair'

LEARNING SUMMARY

Newton's laws of motion

With his three laws of motion, Newton thought that he could describe and predict the movement of every object in the Universe. Like many other 'laws' in physics, they give the right answers much of the time. The paths of planets, moons and satellites are all worked out using Newton's laws.

The ideas behind the three laws have already been used informally in this chapter. This section looks at formal statements of the first and third laws and how they apply to everyday situations.

The first law

AQA A	M2	NICCEA	M1
AQA B	M1	OCR A	A2
EDEXCEL A	M1	OCR B	M1
EDEXCEL B	M1	WJEC	A2

> **KEY POINT**
>
> Newton's first law states that:
> *An object maintains its state of rest or uniform motion unless there is a resultant force acting on it.*

The phrase 'uniform motion' means moving in a straight line at constant speed; i.e. moving at constant velocity.

TEXT	M2
TEXT	M1
TEXT	A2

At first, this seems contrary to everyday experience. Give an object a push and it slows down before coming to rest. This is not 'continuing in a state of uniform motion'. Newton realised that in this case the unseen resistive forces of friction and air resistance together act in opposition to the motion; as there is no longer a driving force after the object has been pushed, there is a resultant backwards-directed force acting on it.

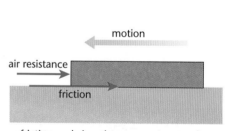

friction and air resistance create a resultant force on this sliding object

Motion without resistive forces is difficult to achieve on Earth: there is always air resistance or friction from a surface that a moving object rests on. The nearest that we can get to modelling motion with no resistive forces is to study motion on ice or an air track.

The converse of Newton's first law is also true:

> if an object is at rest or moving at constant velocity the resultant force on it is zero

The closed figure is a triangle if there are three forces acting; a quadrilateral for four forces, etc.

This means that where there are two forces acting on an object that satisfies these conditions, they must be equal in size and opposite in direction. If there are three or more forces acting then the vector diagram is a closed figure.

A normal reaction force acts on all four wheels; the arrow shown on the diagram represents the sum of these forces.

You have already studied examples of a vehicle moving at constant velocity where the driving force and resistive force are equal in size and act in opposite directions. The diagram shows an example of a situation where three forces sum to zero; it illustrates a vehicle parked on a hill at rest and a vector triangle that shows the forces.

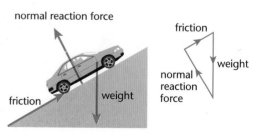

The third law

AQA A	M2	NICCEA	M1
AQA B	M1	OCR A	M1
EDEXCEL A	M1	OCR B	M1
EDEXCEL B	M1	WJEC	M1

The laws were originally written in Latin. The third law is directly translated as 'to every action there is an equal and opposite reaction'. This was widely misinterpreted as meaning that two equal and opposite forces act on the same object, resulting in zero acceleration.

This is both the simplest in its statement and the most misunderstood of the three laws. The statement given here is not a direct translation of the original, but it helps to remove some of the misunderstanding.

> **KEY POINT**
>
> Newton's third law can be stated as:
>
> If object A exerts a force on object B, then B exerts a force equal in size and opposite in direction on A.

According to the third law, forces do not exist individually but in pairs. However, it is important to remember that:

- the forces are of the same type, i.e. both gravitational or electrical
- the forces act on different objects
- the third law applies to all situations.

Some examples

AQA A	M2	NICCEA	M1
AQA B	M1	OCR A	M1
EDEXCEL A	M1	OCR B	M1
EDEXCEL B	M1	WJEC	M1

If you step off a wall the Earth pulls you down towards the ground. According to the third law, you also pull the Earth up with an equal-sized force. So why doesn't the Earth accelerate upwards to meet you instead of the other way round? The answer is it does, but if you apply $F = ma$ to work out the Earth's acceleration, it turns out to be minimal.

Planets pull moons, and according to the third law, moons pull planets with an equal-sized force. Why does the moon go round the planet instead of the other way round?

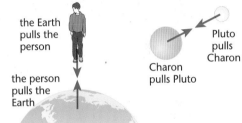

Newton's third law: pairs of forces that are equal in size and opposite in direction.

Pluto and its moon Charon are so close together in terms of their mass that they spin round like a pair of dancers, each of them orbiting a point in space between the planet and its moon.

This type of situation is where the greatest misunderstanding of Newton's third law occurs. Remember, third law pairs of forces act on **different** objects.

Again, the answer is, it does. In fact, they both rotate around a common centre of mass. In the case of the Earth and its Moon, the centre of mass is so close to the centre of the Earth that to all intents and purposes the Moon orbits the Earth.

The diagram shows another pair of forces that are equal in size and opposite in direction. This is an application of the first law – the vase is in a state of rest so there is no resultant force acting on it.

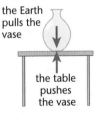

These forces are also equal in size and opposite in direction – an application of Newton's first law, not third.

Progress check

1. A tightrope walker stands in the middle of a rope.
   a. What exerts the upward force on the tightrope walker?
   b. Explain why it is not possible for the rope to be perfectly horizontal.

2. Explain how you can tell from a vector diagram that an object is in a state of rest or uniform motion.

3. Identify the other forces that make up the third law pairs in the diagram of the vase above.

3 The vase pulls the Earth.
The vase pushes the table.
2 The arrows that represent the vectors form a closed figure; i.e. the last arrow finishes where the first one starts.
1 a The tension in the rope.
b A rope can only pull in its own direction. For the tension to have a vertical component, the rope must be inclined to the horizontal.

1.8 Momentum and the second law

After studying this section you should be able to:

- apply the principle of conservation of momentum to collisions in one dimension
- state the relationship between the change of momentum of an object and the resultant force on it
- explain the significance of the impulse of a force

Momentum

AQA A	M2	NICCEA	M1
AQA B	A2	OCR A	A2
EDEXCEL A	M1	OCR B	A2
EDEXCEL B	A2	WJEC	A2

A collision between two objects need not involve physical contact. Imagine two positive charges approaching each other; the repulsive forces could cause them to reverse in direction without actually touching.

Which is more effective at demolishing a brick wall, a 1 kg iron ball moving at 50 m s^{-1} or a 1000 kg iron ball moving at 1 m s^{-1}? Both the mass and the velocity need to be taken into account to answer this question.

Newton realised that what happens to a moving object involved in a collision depends on two factors:

- the mass of the object
- the velocity of the object.

He used the concept of **momentum** to explain the results of collisions between objects.

> **KEY POINT**
>
> momentum = mass × velocity
>
> $$p = mv$$
>
> Momentum is a vector quantity, the direction being the same as that of the velocity. It is measured in N s or kg m s^{-1}; these units are equivalent.

A ten-pin bowling ball has a mass of several kilogrammes, so even at low speeds it has much more momentum than a tennis ball (mass 0.06 kg) travelling as fast as you can throw it.

Everything that moves has momentum and exerts a force on anything that it interacts with. This applies as much to the light that is hitting you at the moment as it does to a collision between two vehicles.

No matter how hard you throw a tennis ball at a garden wall you will not knock the wall down; nor will you be very successful if you use a tennis ball to play ten-pin bowling!

Conservation of momentum

AQA A	M2	NICCEA	A2
AQA B	A2	OCR A	A2
EDEXCEL A	M1	OCR B	A2
EDEXCEL B	A2	WJEC	A2

The phrase 'equal and opposite' is used as a shorthand way of writing 'equal in size and opposite in direction'.

When a force causes an object to change its velocity, there is also a change in momentum. When two objects interact or collide, they exert equal and opposite forces on each other and so the momentum of each one changes. This is illustrated below where the blue ball loses momentum and the black ball gains momentum.

Like the forces they exert on each other during the collision, the changes in momentum of the balls are *equal in size and opposite in direction*.

It follows that:

- the combined momentum of the balls is the same before and after they collide.

This is an example of the **principle of conservation of momentum**.

To satisfy the conditions for conservation of momentum to apply, the only forces acting on the objects must be the ones between the objects themselves.

This is why conservation of momentum is usually demonstrated on an air track or other low friction surface.

> The principle of conservation of momentum states that:
> When two or more objects interact the total momentum remains constant provided that there is no external resultant force.

In the context of the colliding balls, an external resultant force could be due to friction or something hitting one of the balls. Either of these would result in a change in the total momentum.

Different types of interaction

AQA A	M2	NICCEA	A2
AQA B	A2	OCR A	A2
EDEXCEL A	M1	OCR B	A2
EDEXCEL B	A2	WJEC	A2

Objects moving in opposite directions

When two objects approach each other and collide head-on, the result of the collision depends on whether they stick together or whether one or both rebound.

The diagram shows two 'vehicles' approaching each other on an air track. The vehicles join together and move as one.

To make two air track vehicles stick together, one has some Plasticine stuck to it and the other is fitted with a pin that sticks into the Plasticine.

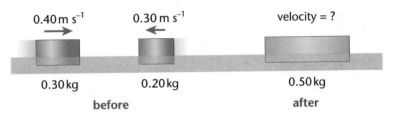

0.40 m s⁻¹   0.30 m s⁻¹   velocity = ?

0.30 kg   0.20 kg   0.50 kg

before   after

The principle of conservation of momentum can be used to work out the combined velocity, v, after the collision.

Taking velocity from left to right as being positive:

total momentum before collision =
0.3 kg × +0.4 m s⁻¹ + 0.2 kg × −0.3 m s⁻¹ = 0.06 N s.

This must equal the momentum after the collision, so 0.5 kg × v = 0.06 N s,

i.e. v = 0.12 m s⁻¹. As this is positive, it follows that the 'vehicle' is moving from left to right.

Rebound

When a light object collides with a heavier one, the light object may rebound, reversing the direction of its momentum. The diagram shows an example of such a collision.

A rebound collision can be modelled on an air track by fitting the vehicles with repelling magnets.

before   after

10 m s⁻¹   0 m s⁻¹ v 3.3 m s⁻¹

2 kg   10 kg

The figures here are rounded. The 3.33 is a rounded value of 3⅓.

Application of the principle of conservation of momentum in this case gives:

2 kg × 10 m s⁻¹ + 10 kg × 0 m s⁻¹ = 2 kg × v + 10 kg × 3.33 m s⁻¹

so 2 kg × v = 20 Ns − 33.3 Ns i.e. v = −6.7 m s⁻¹

The negative value means the object is moving from right to left.

Recoil

In some situations two objects are stationary before they interact, having a combined momentum of zero. After they interact, both objects move off but the combined momentum is still zero. Examples of this include:

- a stationary nucleus undergoes radioactive decay, giving off an alpha particle
- a person steps off a boat onto the quayside
- two ice skaters stand facing each other; then one pushes the other
- two air track vehicles fitted with repelling magnets are held together and then released.

In each case, both objects start moving from rest. For the combined momentum to remain zero, each object must gain the same amount of momentum but these

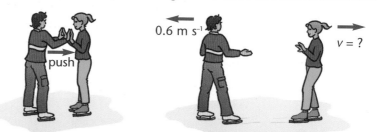

momentum gains must be in *opposite directions*. The ice skater who does the pushing **recoils** as an equal and opposite force is exerted by the one who is pushed.

In the example shown in the above diagram, an 80 kg man pushes a 50 kg child. Their combined momentum is zero both before and after this interaction.

Taking velocity from left to right as positive, this means that:

$$80 \text{ kg} \times -0.6 \text{ m s}^{-1} + 50 \text{ kg} \times v = 0$$

so *v*, the final velocity of the child, is 0.96 m s^{-1} to the right.

Elastic or inelastic

AQA A	M2	OCR A	A2
EDEXCEL B	A2	WJEC	A2
NICCEA	A2		

In all three examples above, both momentum and energy are conserved. This is always the case. A collision where the total kinetic energy before and after the interaction is the same is called an **elastic** collision. Where there is a gain or loss of kinetic energy as a result of the interaction the collision is called **inelastic**.

> in an elastic collision: momentum, kinetic energy and total energy are conserved
> in an inelastic collision: momentum and total energy are conserved, but the amount of kinetic energy changes

Of the examples given, only the second is an elastic collision.

Momentum and the second law

AQA A	M2	NICCEA	M1
AQA B	A2	OCR A	A2
EDEXCEL A	M1	OCR B	A2
EDEXCEL B	A2	WJEC	A2

Newton's second law establishes the relationship between the resultant force acting on an object and the change of momentum that it causes. Important results that follow from the second law include $F = ma$, studied in force and acceleration, and the **impulse of a force**, to be met later in this section.

> **KEY POINT**
>
> Newton's second law states that:
>
> The rate of change of momentum of an object is proportional to the resultant force acting on it and acts in the direction of the resultant force.
> $$\Delta p / \Delta t \propto F$$

The definition of the newton (see force and acceleration) fixes the proportionality constant at 1. This enables the second law to be written as:

> **Force = rate of change of momentum**
> $F = \Delta p/\Delta t$
>
> KEY POINT

In this form, the second law is useful for working out the force in situations such as jet and rocket propulsion where the change in momentum each second can be calculated easily.

A rocket carries its own oxygen supply, so that it can fire the engines when it is travelling in a vacuum. A spacecraft travelling where it is not affected by any gravitational fields or resistive forces maintains a constant velocity, so it only needs to fire the rocket engines to change speed or direction.

The principle of rocket propulsion can be seen by blowing up a balloon and letting it go; the air squashed out of the neck of the balloon gains momentum as it leaves, causing the balloon and the air that remains inside to gain momentum **at the same rate but in the opposite direction**. The diagram shows the principle of rocket propulsion.

Jet and rocket engines use the same principle; hot gases are ejected from the back of the engine to provide a force in the forwards direction. The difference is that jet engines are designed to operate within the Earth's atmosphere, so they can take oxygen from their surroundings.

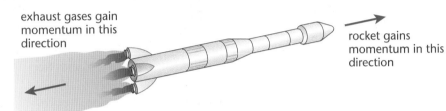

exhaust gases gain momentum in this direction

rocket gains momentum in this direction

conservation of momentum and rocket propulsion

The unit kg m s⁻² is equivalent to the N, so a rate of change of momentum of 4 500 000 kg m s⁻² is another way of stating: a force of 4 500 000 N.

In this example, the exhaust gases are emitted at a rate of 3000 kg s⁻¹. The speed of the exhaust gases relative to the rocket is 1500 m s⁻¹. The rate of change of momentum of these gases is therefore 4 500 000 kg m s⁻², i.e. 4 500 000 kg m s⁻¹ each second. This is the size of the force on both the gases and the rocket.

Impulse

AQA B	A2	NICCEA	M1
EDEXCEL A	M1	OCR B	A2
EDEXCEL B	A2		

A modern jet engine is capable of producing a force of 1 MN (1×10^6 N). Four of these engines are used to accelerate a jumbo jet to the speed required for take-off.

Jet and rocket engines are often working over a long period of time, so the second law written as **force = rate of change of momentum** can be readily applied when the time for which the force acts is not a fixed quantity. A manufacturer of aircraft engines, for example, uses the maximum force required to design the engine to achieve the required rate of output of exhaust gases.

In situations where the time for which the force acts is known, or is short, the concept of **impulse** can be more readily applied. The impulse of a force is the change in momentum that it causes. Using the concept of impulse, the second law can be written as:

> **impulse** = force × time
> = change in momentum
> Impulse, like momentum, is measured in the equivalent units of N s or kg m s⁻¹.
>
> KEY POINT

When a child is taught how to catch a hard ball, he is shown how to pull back the hands while catching the ball. This extends the time required to stop the ball, so a smaller force is exerted on the hands.

When tennis players or snooker players hit a ball, they are aware that the effect of the force that they exert depends on the time of contact with the ball as well as the size of the force. By maintaining contact with the ball over a longer period of time, a greater change in momentum can be caused by a given force.

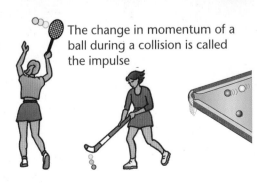

The change in momentum of a ball during a collision is called the impulse

In the example of a ball hitting a wall and rebounding, the force is not constant. It becomes greater as the ball becomes more compressed and decreases again as the ball returns to its original shape. The diagram below shows how the force on a ball changes during a collision with a wall.

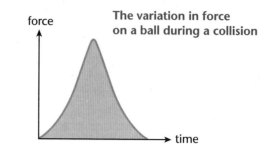

force

The variation in force on a ball during a collision

time

> Remember, the area between the curve on a graph and the *x*-axis represents the product of the quantities plotted on the *x*- and *y*- axes.

The impulse, or change in momentum of the ball, is represented by the area between the curve and the time axis, shown shaded on the graph.

> In practice sessions, long-jumpers and high-jumpers often land on a bed made of pieces of foam rubber. This gives a long 'stopping' time, so there is no chance of physical injury due to the impulse as they land.

Although the concept of impulse is often used to describe the effect of a force that acts over a short time, it can be applied to any interaction between objects. When a pole-vaulter lands on the ground, her momentum is typically 400 to 500 N s. This is the size of the impulse needed to stop her. On a hard concrete surface she could be brought to rest in a tiny fraction of a second, needing a back-breaking force. As **impulse = force × time**, the same change in momentum can be achieved by a smaller force acting over a longer time.

The cushions lengthen the time of the impulse, so a small force is exerted.

Progress check

1 In an air track collision, a 0.10 kg vehicle travelling at 0.20 m s⁻¹ from left to right approaches a 0.30 kg vehicle travelling at 0.30 m s⁻¹ in the opposite direction. After the head-on collision, the vehicles stick together. Calculate their combined velocity.

2 A rocket burns fuel at the rate of 3000 kg each second. The exhaust gases have a speed, relative to the rocket, of 2200 m s⁻¹.
   a Calculate the rate of change of momentum of the exhaust gases.
   b The mass of the rocket is 500 000 kg. Calculate the acceleration while the fuel is being burned. Assume that no other forces act on the rocket.
   c Suggest why the acceleration of the rocket increases as more fuel is burned.

3 A tennis ball with no initial horizontal velocity leaves a tennis racquet with a horizontal speed of 58 m s⁻¹. The mass of the ball is 0.060 kg.
   a Calculate the impulse given to the ball.
   b If the time of contact with the racquet is 0.025 s, calculate the average force exerted on the ball.

b 139 N
3 a 3.48 N s
c The mass of the rocket decreases as more fuel is burned.
b 13.2 m s⁻²
2 a 6 600 000 kg m s⁻²
1 0.175 m s⁻¹ from right to left.

1.9 Energy to work

After studying this section you should be able to:

- identify situations where a force is working
- state and use the relationships between force, distance, energy, work, power and time
- calculate the efficiency of an energy transfer

Working

AQA A	M2	NICCEA	A2
AQA B	A2	OCR A	M1
EDEXCEL A	M1	OCR B	M2
EDEXCEL B	M1	WJEC	A2

Every event requires **work** to make it happen. Even if you relax and try to do nothing, your body is working just to keep you alive. Any force that causes movement is doing work. Pushing a supermarket trolley is working, as is throwing or kicking a ball. However, holding some weights above your head is not working; it may cause your arms to ache, but the force on the weights is not causing any movement!

How much work a force does depends on:
- the size of the force
- the direction of movement
- the distance that an object moves.

The phrase, 'distance moved in the direction of the force', is used here instead of 'displacement' to emphasise that the force and displacement it causes are measured in the same direction.

> **KEY POINT**
>
> The work done by a force that causes movement is defined as:
>
> work = average force × distance moved in the direction of the force
>
> $$W = F \times s$$
>
> Work is measured in joules (J) where 1 joule is the work done when a force of 1 N moves its point of application 1 m in the direction of the force.

Here are some examples of forces that are working and forces that are not working.

The shelf is not moving, so no work is being done.

The tension in the string is not working as there is no movement **in the direction of the force**.

A pylon that supports electricity transmission cables is not working, but a wind that causes the cables to move is!

This force causes movement **in the direction of the force**.

The horizontal component of the force on the log is doing the work here.

In circular motion, the direction of the velocity is along a tangent to the circle.

In the case of the stone being whirled in a horizontal circle, the stone's velocity is always at **right angles** to the force that maintains the motion, so the force is not working.

Although the force pulling the log and the movement it causes are not in the same direction, the force on the log has a component in the same direction as the log's movement. In calculating the work done the horizontal component of the force is multiplied by the horizontal distance moved by the log.

Work and energy transfer

AQA A	M2	NICCEA	A2
AQA B	A2	OCR A	M1
EDEXCEL A	M1	OCR B	M2
EDEXCEL B	M1	WJEC	A2

The physical processes of **working** and **energy transfer** are inseparable. Work requires an energy source, for example, a fuel, an electric cell or a wound-up spring. As the work is done, energy leaves the source and ends up elsewhere, this is what is meant by **energy transfer**.

> An object has energy if it can exert a force that causes movement of the point of application of the force.

> **KEY POINT**
>
> Energy is the ability to do work.
> The energy transfer from a source is equal to the amount of work done.
> Like work, energy is measured in joules.

Here are some examples of everyday energy transfers.

A bus accelerates away from a bus stop and then maintains a steady speed

> A common error is to describe the energy transfer of a vehicle moving at a constant speed as 'energy from the fuel and oxygen is transferred to kinetic energy'.

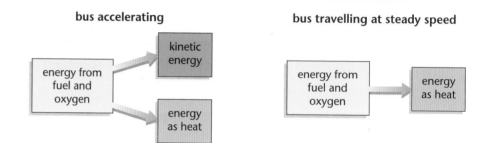

As the bus speeds up, some of the energy from the fuel and oxygen is transferred to the **kinetic energy** of the bus, i.e. the energy it has due to its movement. The rest of the energy is transferred to the surroundings. This takes place in a number of ways: the exhaust gases transfer energy as heat directly into the surrounding air and some energy is also transferred to heat wherever there are resistive forces.

Once the bus is maintaining a steady speed there is no increase in its kinetic energy, so all the energy transferred from the fuel and oxygen is going to the surroundings.

An electric motor lifts a load at a steady speed

> The phrase 'gravitational potential energy' is often abbreviated to 'gpe' or just 'potential energy', although strictly 'potential energy' can be any form of stored energy, for example in a spring or a fuel.

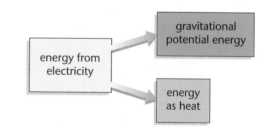

The load gains **gravitational potential energy** as it is lifted, i.e. the energy it has due to its position above the Earth's surface. All the movement is at a constant speed, so there is no change in the kinetic energy of the system. As with the bus travelling at a steady speed, work done against resistive forces causes heating, and this energy is transferred as heat to the surroundings.

Energy and a filament lamp

AQA A	M2	OCR A	A2
EDEXCEL A	M1	WJEC	A2

When a filament lamp is switched on, some energy is absorbed by the filament as it reaches its operating temperature. This takes a fraction of a second. Some energy is absorbed by the glass envelope and the gas inside. These take a little longer to reach a steady temperature, perhaps a few seconds.

Once the lamp has reached its steady state, of the 60 J of energy passing into the lamp each second from the electricity supply, typically 3 J passes out as light. This is shown in the diagram. The Sankey diagram gives a visual indication of the relative proportions of energy transferred into a desirable output and wasted.

Energy flow through filament lamp

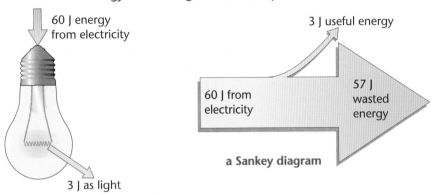

60 J energy from electricity

3 J as light

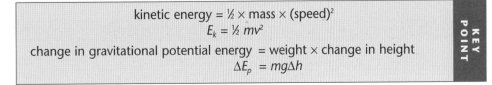

3 J useful energy

60 J from electricity

57 J wasted energy

a Sankey diagram

The 'wasted' energy is transferred to the surroundings in two main ways:
• from the glass envelope to the surrounding air by conduction and convection
• from all parts of the lamp as non-visible electromagnetic radiation (mainly infra-red).

The light and the infra-red radiation both cause heating when they are absorbed, so all the energy input to the lamp ends up as heat!

These examples illustrate some important points about energy transfer:
• energy due to movement is called kinetic energy
• energy due to position above the Earth's surface is called gravitational potential energy, often abbreviated to potential energy
• work usually involves some transfer of energy to the surroundings.

Conservation of energy

AQA A	M2	NICCEA	M1
AQA B	M1	OCR A	M1
EDEXCEL A	M1	OCR B	M2
EDEXCEL B	M1	WJEC	A2

The examples of energy transfer identify some ways in which energy flows into and out of a process where work is done. Like mass and charge, energy is a conserved quantity.

> The principle of conservation of energy states that:
> energy cannot be created or destroyed.
>
> **KEY POINT**

This simple statement means that energy is never **used up** although often the energy output from a working process is in a form which is difficult to harness to do more work. The energy output from a filament lamp, for example, is effectively **lost** as heat in the surroundings. Like the energy from the bus, it raises the temperature of its surroundings by such a small amount that makes it difficult to recover.

In many energy transfer processes the amount of energy at each stage cannot easily be quantified; it is difficult to put a figure on the total kinetic and potential energy of the moving parts in an engine for example. There are simple formulas for calculating the kinetic energy and gravitational potential energy of individual objects.

> kinetic energy = ½ × mass × (speed)²
> $E_k = \frac{1}{2} mv^2$
> change in gravitational potential energy = weight × change in height
> $\Delta E_p = mg\Delta h$
>
> **KEY POINT**

The formula for calculating gravitational potential energy gives the *change* in energy because ground level varies and so does not provide an accurate reference point for zero energy. This is not the case with kinetic energy as no movement relative to the Earth's surface is the same the world over.

The principle of conservation of energy applies to all energy transfers.

When a streamlined object is falling freely in the atmosphere at a low speed, the resistive forces are small and so little energy is transferred to the surroundings. In this case the principle of conservation of energy can be applied to the transfer of gravitational potential energy to kinetic energy. When a parachutist is falling at a constant speed conservation of energy still applies, but the energy transfer is from gravitational potential energy to heat in the atmosphere.

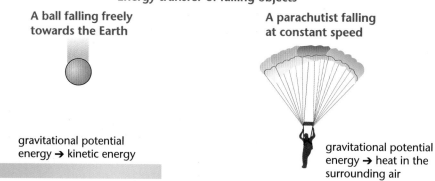

Energy transfer of falling objects

A ball falling freely towards the Earth

A parachutist falling at constant speed

gravitational potential energy → kinetic energy

gravitational potential energy → heat in the surrounding air

Efficiency

AQA B	A2	NICCEA	A2
EDEXCEL A	M2	OCR A	A2
EDEXCEL B	M1	WJEC	A2

A filament lamp is a good example of an everyday device that does not do its job very well. Of the 60 J of energy input to the lamp each second, only 3 J is output as light. It has a very low efficiency!

An energy-efficient lamp has an efficiency of 0.25 (25%), compared to 0.05 (5%) for a filament lamp.

Fluorescent lamps are much more **efficient** at doing the same job, but until recently they were only available in the form of long tubes. Modern energy-efficient lamps are fluorescent lamps that are designed as direct replacements for filament lamps. Although their running costs are much lower and they last much longer, the initial cost is higher. To achieve the same light output, an energy-efficient lamp needs only 20% of the energy input from electricity that a filament lamp needs. It is more efficient at doing the same job.

> Efficiency is defined as:
>
> useful energy output ÷ total energy input
>
> It is usually expressed as a fraction or decimal, which can be multiplied by 100 to give a percentage efficiency.
>
> **KEY POINT**

Power

AQA A	M2	NICCEA	A2
AQA B	A2	OCR A	M1
EDEXCEL A	M1	OCR B	M2
EDEXCEL B	M1	WJEC	A2

In choosing or designing a machine to do a particular job, important factors to consider include the **power input** and the **power output**. You would expect a hairdryer with a high power output to dry your hair faster than one with a low power output, but a vacuum cleaner with a high power input may not clean the carpet any faster than one with a lower power input, it may just be less efficient!

Do not confuse power with speed. A double-decker bus may be slower than many cars, but it can transport sixty passengers a given distance much faster than any car!

Steam trains have a very high power input, but a very low efficiency at transferring this to output power.

Typically, a car engine has an output power of 70 kW, compared to 1.5 kW for a hairdryer and 2 W for a clock.

Power is a measure of the work done or energy transferred each second.

> **KEY POINT**
>
> Power is the rate of working or energy transfer.
>
> Average power can be calculated as work done ÷ time taken, $P = \Delta W \div \Delta t$
>
> Power is measured in watts (W), where 1 watt is a rate of working of 1 joule per second (J s^{-1})

Power and velocity

AQA A	M2	NICCEA	A2
AQA B	A2	OCR A	M1
EDEXCEL A	M1	OCR B	M2
EDEXCEL B	M1		

The expression $(F \times s) \div t$, representing **work done ÷ time taken** is also equal to **force × velocity**. This gives an alternative way of calculating power in situations where the force and speed in the direction of the force (the component of the velocity in that direction) are known:

> **KEY POINT**
>
> Power = force × velocity
>
> $P = Fv$

This expression can be applied to the motive power of a vehicle, where F is the driving force and v is the speed of the vehicle.

For example, if a force of 600 N is used to tow a trailer at a speed of 12 m s^{-1}, then the motive power required is 600 N × 12 m s^{-1} = 7200 W.

Power and efficiency

AQA B	A2	EDEXCEL B	M1
EDEXCEL A	M2	NICCEA	A2

The output power of a device depends on both the input power and the efficiency.

Where the power input and output of a machine are known, it is not necessary to calculate the work input and output in order to calculate the efficiency. The input and output powers can be used directly. This gives the alternative relationship for efficiency:

> **KEY POINT**
>
> efficiency = useful power output ÷ total power input

Progress check

1 A 240 N force is used to drag a 50 kg mass a distance of 5.0 m up a slope (see diagram). The mass moves through a vertical height of 1.8 m. Take $g = 10$ m s^{-2}. Calculate:

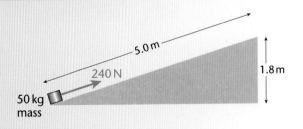

   a   The work done on the mass.
   b   The gravitational potential energy gained by the mass.
   c   The efficiency of this way of lifting the mass through a vertical height of 1.8 m.
   d   The time taken to drag the mass up the slope by a motor that has an output power of 150 W.

2 A car pulls a trailer with a force of 750 N, transmitted through the tow bar. The tow bar is inclined at 20° to the horizontal. Calculate the work done by the car in pulling the trailer through a horizontal distance of 200 m.

3 A car has a maximum speed of 65 m s^{-1}. At this speed, the motive power is 90 kW. Calculate the value of the driving force.

3 1385 N
2 1.4×10^5 J
d 8.0 s
c 0.75
b 900 J
1 a 1200 J

1.10 Drag and terminal velocity

After studying this section you should be able to:

- *explain the difference between laminar and turbulent flow*
- *calculate the drag force acting on a vehicle and explain how it varies with the speed of the vehicle*
- *explain why objects reach a terminal velocity*

LEARNING SUMMARY

Streamlines and turbulence

EDEXCEL A ▸ M1   OCR A ▸ M1
EDEXCEL B ▸ M2

In a wind tunnel the air moves over a stationary object such as a car body to model the movement of the car through the air. This is a valid model, as it is the relative speed of the air and the object that determines the pattern of flow.

To be as fuel-efficient as possible, a car, train or aircraft needs to be designed to minimise the resistive forces acting on it. To investigate the resistive or **drag** forces due to movement through the air, a car body is placed in a **wind tunnel**.
Here the car body is held stationary and the air moves around it; vapour trails show the air flow over the car body.

The diagram below shows the air flow at a low speed over a car body. The flow is said to be **laminar** or **streamlined**.

In laminar flow:

- the air moves in layers, with the layer of air next to the car body being stationary and the velocity of the layers increasing away from the car body
- particles passing the same point do so at the same velocity, so the flow is regular
- the drag force is caused by the resistance of the air to layers sliding past each other
- more **viscous** air has a greater resistance to relative motion and exerts a bigger drag force
- the drag force is proportional to the speed of the car relative to the air.

Viscosity is a measure of the resistance of a fluid to flowing. Syrup and tar are viscous fluids; hydrogen has a very low viscosity.

The air flow over a car travelling at low speed   turbulence occurs at higher speeds

As the speed of the air passing over the car body is increased, the flow pattern changes from laminar to **turbulent**. This is shown in the diagram above.

The changeover from laminar flow to turbulent occurs at a speed known as the **critical velocity**. Turbulent flow causes much more drag than laminar flow.

In turbulent flow:

- the air flow is disordered and irregular
- the drag force depends on the **density** of the air and not the viscosity
- the drag force is proportional to the $(speed)^2$ of the car relative to the air.

Terminal velocity

AQA A	M1	OCR A	M1
AQA B	M1	OCR B	M2
EDEXCEL B	M2	WJEC	M1

If an object is falling through a fluid, e.g. air or water, as its speed increases the drag on it will also increase. Eventually a speed is reached where the upward force will equal the weight of the object. As there is no net force on the object the acceleration will be zero. The object will fall at a constant velocity. This velocity is called the **terminal velocity**.

Similarly for a car, or other vehicle, when it reaches a speed where the sum of all the forward forces, i.e. the thrust = the sum of all the resistive forces, i.e. the drag, the vehicle will move at a terminal velocity.

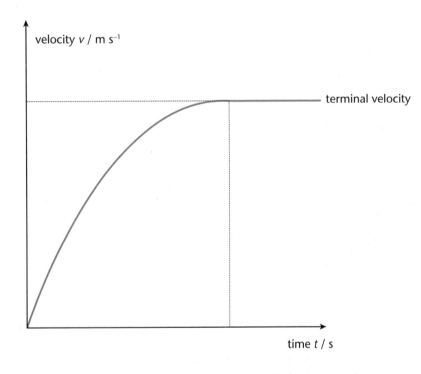

Progress check

1 State **two** differences between laminar flow and turbulent flow of air over an object such as a car body.

2 Suggest why it is dangerous for an aircraft to attempt to take off when there is a layer of ice on its wings.

3 The aerofoils that form the wings of an aircraft have a lift coefficient, $C_L = 50$. The surface area of the underside of each wing is 10 m^2 and the aircraft leaves the ground when its speed through the air is 60 m s^{-1}. Calculate the lift force required for take-off if the density of the air is 1.1 kg m^{-3}.

3 1.98×10^6 N

2 the ice increases the turbulence, reducing the lift.

1 any two from: laminar flow is regular; in laminar flow layers of air slide over each other; in laminar flow the drag force is due to resistance of the layers of air to sliding over each other; in laminar flow the drag force is proportional to the relative speed of the object and the air; in laminar flow the drag is proportional to the viscosity of the air.

Sample question and model answer

In an emergency stop, a driver applies the brakes in a car travelling at 24 m s⁻¹. The car stops after travelling 19.2 m while braking.

(a) Calculate the average acceleration of the car during braking. [3]

The final velocity, v, is implicit in the question, although not clearly stated. As the car brakes to a halt, the final velocity must be 0.

The initial velocity, u, final velocity, v, and displacement, s, are known.
The unknown quantity is the acceleration, a.
The appropriate equation is $v^2 = u^2 + 2as$ 1 mark
Rearrange to give $a = (v^2 - u^2)/2s$ 1 mark
$$= (0 - 24^2)/(2 \times 19.2)$$

Note that the acceleration is negative, as the speed in the direction of the velocity is decreasing.

$$= -15 \text{ m s}^{-2}$$ 1 mark

(b) Sketch a graph that shows how the speed of the car changes during braking, assuming that the acceleration is uniform. Mark the value of the intercept on the time axis.
Show on your graph the feature that represents the distance travelled during braking. [3]

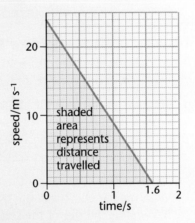

The graph shows a straight line from 24 m s⁻¹ to 0 m s⁻¹   1 mark
The intercept on the time axis is 1.6 s   1 mark
The area between the graph line and the time axis represents the distance travelled.   1 mark

(c) Calculate the size of the force required to decelerate the driver, of mass 75 kg, at the same rate as the car. [3]

Again, the negative sign shows that the direction of the force is opposite to the direction of motion.

force = mass × acceleration   1 mark
$$= 75 \text{ kg} \times -15 \text{ m s}^{-2}$$ 1 mark
$$= -1125 \text{ N}$$ 1 mark

(d) Suggest what is likely to happen to a driver who is not wearing a seat belt. The car does not have air bags. [2]

The cue word 'suggest' indicates that you are not meant to have studied this example as part of your course, but should be able to reason it from your understanding of forces and motion.

The driver will carry on moving (1 mark) as there is nothing to stop her/him until (s)he hits the steering wheel and windscreen (1 mark).

Practice examination questions

Take the value of free-fall acceleration, g to be 10 m s^{-2}.

1 An aircraft has a total mass, including fuel and passengers, of 70 000 kg. Its take-off speed is 60 m s^{-1} and it needs to reach that speed before the end of the runway, which is 1500 m long.

(a) Calculate the minimum acceleration of the aircraft. [3]

(b) Calculate the average force needed to achieve this acceleration. [3]

(c) Explain how the resultant force on the aircraft is likely to change during take-off. [2]

2 A car pulls a trailer with a force of 150 N. This is shown in the diagram.

150 N

(a) According to Newton's third law, forces exist in pairs.
(i) What is the other force that makes up the pair? [1]
(ii) Write down the size and direction of this force. [1]

(b) The mass of the trailer is 190 kg. As it sets off from rest, there is a resistive force of 20 N.
(i) Calculate the size and direction of the resultant force on the trailer. [1]
(ii) Calculate the initial acceleration of the trailer. [3]

(c) The force from the car on the trailer is maintained at 150 N. When the car and trailer are travelling at constant speed:
(i) what is the size of the resultant force on the trailer? [1]
(ii) write down the size and direction of the resistive force on the trailer. [1]

3 The diagram shows the forces acting on a child on a playground slide. Air resistance is negligible.

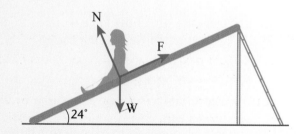

(a) The mass of the child is 40 kg.
(i) Calculate the size of the force W. [1]
(ii) Describe the force that, along with W, makes up the 'equal and opposite pair' of forces described by Newton's third law. [1]

(b) The size of the friction force is 90 N.
(i) Calculate the component of W parallel to the slide. [2]
(ii) Calculate the acceleration of the child. [3]

(c) The slide is 5.5 m long. If the child maintains this acceleration, calculate her speed at the bottom. [3]

Practice examination questions (continued)

4 The diagram shows two 'vehicles' on an air track approaching each other. After impact they stick together.

0.45 kg   0.30 kg

The speed of the heavier vehicle is 0.60 m s^{-1}.

(a) If the lighter vehicle is travelling at 0.42 m s^{-1}, calculate the velocity of the vehicles after the collision. [3]

(b) If both vehicles are stationary after the collision, what is the speed of the lighter vehicle before the collision? [2]

5 The diagram shows a neutron of mass 1.7×10^{-27} kg about to collide inelastically with a stationary uranium nucleus of mass 4.0×10^{-25} kg. During the collision, the neutron will be absorbed by the uranium nucleus.

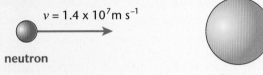

$v = 1.4 \times 10^{7}$ m s^{-1}

neutron

uranium nucleus

(a) Calculate the velocity of the uranium nucleus immediately after the neutron has been absorbed. [3]

(b) Collisions between neutrons and uranium nuclei can also be elastic. State, and explain briefly, how the speed of the uranium nucleus after impact would be different in the case of an elastic collision.
Do not perform any further calculations. [3]

(c) Using the data at the beginning of the question, calculate the kinetic energy of the neutron before it collides with the uranium nucleus. [3]

AEB AS-A PHYS 6 Jan 1999 Q 2

6 An astronaut on a space walk pushes against the side of the spacecraft.

(a) Explain how this causes the velocity of both the astronaut and the spacecraft to change. [2]

(b) The astronaut can change his velocity by firing jets of nitrogen gas. Explain how this changes the velocity of the astronaut. [2]

7 A ball of mass 0.080 kg hits a wall at a speed of 32 m s^{-1} and rebounds at a speed of 22 m s^{-1}.

(a) Calculate the change in momentum of the ball. [2]

(b) If the impact between the ball and the wall lasts for 0.15 s, calculate the average force on the ball. [3]

(c) Explain whether momentum is conserved in this interaction. [2]

Practice examination questions *(continued)*

8 A crane lifts a 3 tonne (3000 kg) load through a vertical height of 18 m in 24 s. Calculate:

    (a) the work done on the load   [3]

    (b) the gain in gravitational potential energy of the load   [1]

    (c) the power output of the crane   [2]

    (d) the power input to the crane if the efficiency is 0.45.   [2]

9 In a hydroelectric power station, water falls at the rate of 2.0×10^5 kg s^{-1} through a vertical height of 215 m before driving a turbine.

    (a) Calculate the loss in gravitational potential energy of the water each second.   [3]

    (b) Assuming that all this energy is transferred to kinetic energy of the water, calculate the speed of the water as it enters the turbines.   [3]

    (c) The water leaves the turbines at a speed of 8 m s^{-1}. Calculate the maximum input power to the turbines.   [2]

    (d) The electrical output is 250 MW. Calculate the efficiency of this method of transferring gravitational potential energy to electricity.   [2]

10 The diagram shows a child's toy. The spring is compressed a distance of 0.15 m using a force of 25 N. When the release mechanism is operated the ball rises into the air.

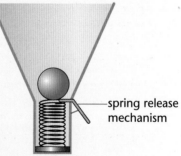

spring release mechanism

    (a) Calculate the energy stored in the spring.   [2]

    (b) The mass of the ball is 0.020 kg. Calculate:

        (i)   the maximum speed of the ball.   [3]

        (ii) the maximum vertical distance that the ball travels.   [3]

Chapter 2

Electricity

The following topics are covered in this chapter:

- *Ohm's law, resistance and resistivity*
- *Energy transfer in circuits*
- *Resistance in circuits*
- *Alternating current and the oscilloscope*
- *Electromagnetism*

2.1 Ohm's Law, resistance and resistivity

After studying this section you should be able to:

- recall and use the formula for resistance
- recall and use the formula for resistivity
- state Ohm's law and describe its limitations
- sketch the current–voltage characteristics of a filament lamp, a wire at constant temperature and a diode

LEARNING SUMMARY

Resistance

AQA A	M3	NICCEA	M1
AQA B	M1	OCR A	M2
EDEXCEL A	M2	OCR B	M1
EDEXCEL B	M1	WJEC	M2

Electromagnetism, heating and electrolysis are all due to currents passing in circuits. Our everyday appliances rely on these three effects of a current.

What determines the current in a component or a circuit? The answers include the voltage; which is a measure of the energy available to move the charge round the circuit, and how freely the charge can flow; which depends on the structure of the conducting material.

Resistance is a measure of the opposition to current. The greater the resistance of a component, the smaller the current that passes for a given voltage.

Just as different objects can weigh different amounts or the same amount, according to their size and the material they are made of, different components can have the same resistance or different resistances. However, if you change the shape of a piece of material its weight does not change but its resistance does.

1 Ω is equivalent to 1 V A^{-1}, so the voltage across a 1 Ω resistor is 1 V for each amp of current passing in it.

> Resistance is defined and calculated using the formula:
> $$\text{Resistance} = \text{voltage} \div \text{current} \quad \text{or} \quad R = V \div I$$
> The unit of resistance is the ohm (Ω).
>
> **KEY POINT**

Different components

AQA A	M3	NICCEA	M1
AQA B	M1	OCR A	M2
EDEXCEL A	M2	OCR B	M1
EDEXCEL B	M1, M2	WJEC	M2

The same piece of equipment is normally used as both a variable resistor and a potential divider. When used as a variable resistor, connections are made to the slider and one of the fixed terminals. All three terminals are used in a potential divider circuit.

There are three key variables that affect the current passing in a conductor. They are:
- the voltage
- the material that the conductor is made from
- the physical dimensions of the conductor.

The diagram shows how a **potential divider** can be used to investigate how the current passing in a component depends on the potential difference, or voltage, across it. Movement of the sliding contact enables the voltage to be varied from zero up to the maximum of the supply voltage.

This section considers some other factors that can determine the size of the current.

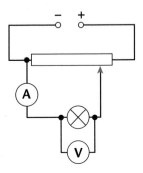

The graphs represent typical results obtained for a metal wire kept at a constant temperature, a filament lamp and a diode.

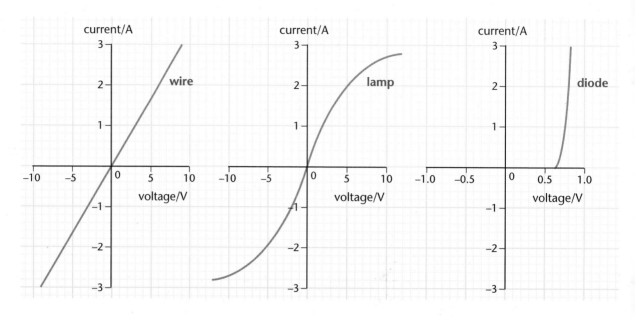

> **KEY POINT**
>
> The current–voltage graph for the **wire** is a straight line through the origin, showing that the *current is directly proportional to the voltage.*

A **filament lamp** is an everyday example of a metallic conductor that does change in temperature during use.

In a short time after it is switched on, the temperature of the filament increases from room temperature to about 2000°C. The current–voltage graph shows how this change in temperature affects the relationship between these quantities. Like the one for the wire at constant temperature, this graph is symmetrical. Both the wire and the lamp filament show the same pattern of behaviour no matter which way round the voltage is applied.

Calculations based on this graph show that **as the current in the filament increases, so does its resistance**. This is due to the higher temperature causing an increase in the amplitude of the lattice vibrations, which in turn increases the frequency of the collisions that impede the electron flow.

The graph showing the characteristics of the **diode** also shows the property that makes diodes different from other components; they only allow current to pass in one direction.

The arrow shows the direction in which the diode allows current to pass.

This direction is shown by the arrow on the diode symbol. With the voltage in the reverse direction, no current passes. This property makes diodes useful in rectification; the process in which a direct current is obtained from an alternating one.

Other diodes in common use include the light-emitting diode (l.e.d.); used as an indicator and the laser diode; used to read information from compact discs.

Even with the voltage in the forwards direction, no current passes until a certain minimum voltage has been reached. The minimum voltage depends on the material that the diode is made from, and is about 0.6 V for a general purpose silicon diode. Once this has been achieved, small increases in voltage cause relatively large increases in current as the resistance of the diode decreases.

Resistivity

AQA A	M3	NICCEA	M1
AQA B	M1	OCR A	M2
EDEXCEL A	M2	OCR B	M1
EDEXCEL B	M2	WJEC	M2

Resistors used in electrical and electronic circuitry are made in a variety of ways. Some are made from a precise length of wire wound on a ceramic base. These tend to be expensive and their use is limited to applications where there is a large current. Resistors used in low-current electronic devices are made from a thin film of conducting material coated onto a small ceramic tube.

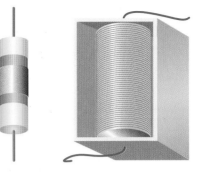

The large wire-wound resistor is able to dissipate heat at the rate of 10 W. This would burn out the small carbon-film resistor.

What factors determine the resistance of either type of resistor?
- The physical dimensions.
  Long wires have more resistance than short wires, so the longer the sample of material, the greater the resistance.

- Cross-sectional area.
  Doubling the cross-sectional area of a sample also doubles the number of charge-carriers available to carry the current, halving the resistance.

> **The resistance of a sample of material is directly proportional to its length and inversely proportional to its cross-sectional area.**
> This statement can be written as $R \propto l/A$.
>
> **KEY POINT**

Good conductors have a low resistivity; that of copper being $1.7 \times 10^{-8}\ \Omega\,\text{m}$. The resistivity of carbon is about one thousand times as great, $3.0 \times 10^{-5}\ \Omega\,\text{m}$.

The other factor that determines the resistance is the material that the resistor is made of; its resistive properties are measured by its **resistivity**, symbol ρ. When this is taken into account the formula becomes $R = \dfrac{\rho l}{A}$. This formula can be used to calculate the resistance when the dimensions and material of a resistor are known or to calculate the dimensions required to give a particular value of resistance.

Resistance is a property of an individual component in a circuit, but resistivity is a property of a material. It is measured in units of $\Omega\,\text{m}$.

Progress check

1 Use the resistance formula to work out the following:
  a The voltage across a lamp filament if the current passing is 2.4 A and the filament resistance is 7.3 Ω.
  b The resistance of an infra-red heater if the current is 9.5 A when it operates from the mains 240 V supply.
  c The current passing in a 6.8 Ω resistor when it is connected to a 9 V battery.

2 Use the resistivity formula to work out the following:
  a The resistivity of silicon. A 1 mm cube of silicon has a resistance of $2.20 \times 10^{6}\ \Omega$ when measured across opposite faces.
  b The length of carbon rod required to make a 3.9 Ω resistor. The resistivity of carbon is $3.00 \times 10^{-5}\ \Omega\,\text{m}$ and the rod has a cross-sectional area of $1.26 \times 10^{-6}\ \text{m}^2$.

3 An electrical heating element is to be made from nichrome wire. The current in the heating element is to be 5.00 A when the voltage across it is 12.0 V.
  a Calculate the resistance of the heating element.
    The resistivity of nichrome is $1.10 \times 10^{-6}\ \Omega\,\text{m}$.
    The cross-sectional area of the wire used is $7.85 \times 10^{-7}\ \text{m}^2$.
  b Calculate the length of wire required.

Note that
$1\ \text{mm} = 1 \times 10^{-3}\ \text{m}$.

First re-arrange the resistivity equation to make *l* the subject.

1 **a** 17.5 V   **b** 25.3 Ω **c** 1.32 A
2 **a** 2200 Ω m   **b** 0.164 m
3 **a** 2.4 Ω **b** 1.71 m

2.2 Energy transfer in circuits

After studying this section you should be able to:

- recall and use the relationship between charge flow and current
- recall and use the relationship between potential difference, energy transfer and charge
- explain the difference between electromotive force and potential difference
- use the relationships between electrical power, current, voltage and resistance

LEARNING SUMMARY

Charge and current

AQA A	M3	NICCEA	M1
AQA B	M1	OCR A	M2
EDEXCEL A	M2	OCR B	M1
EDEXCEL B	M1	WJEC	M2

In a metal, current is due to the movement of free electrons from negative to positive.

In a conducting non-metal, positive and negative ions flow in opposite directions.

To conduct electricity, a material must have charged particles that can move. In metals, these are negatively-charged **free electrons**; while both positively-charged and negatively-charged **ions** are responsible for conduction in gases and non-metallic liquids.

- An overall flow of charge in a particular direction is an **electric current**.
- Current = rate of flow of charge.

The relationship between charge flow and current can be written as $I = \Delta q / \Delta t$, where Δq is the amount of charge that flows in a short time Δt.

> This relationship is used to define the size of the unit of charge, the coulomb is the charge that flows past a point when a steady current of 1 A passes for 1 second.
>
> **KEY POINT**

Current and drift velocity

AQA B	M1	NICCEA	M1
EDEXCEL A	M2	WJEC	M2

The free electrons in a non-conducting metal are in constant motion. The motion of any one electron is random in both speed and direction, and constantly subject to change due to collisions, but a typical speed in a metal at room temperature is of the order of 1×10^6 m s^{-1}.

The random motion of a free electron

drift velocity is slow compared to the random motion of the electrons

Applying an electric force in the form of a voltage does not cause the free electrons to suddenly dash for the positive terminal. The random motion continues, but in addition the body of electrons moves at low speed in the direction negative to positive.

The overall speed of the body of electrons is called the **drift velocity**.
There are three key variables that affect the drift velocity:

- current
- charge carrier concentration, n, defined as the number of charge carriers (free electrons in the case of a metal) per unit volume
- the cross-sectional area of the specimen, A.

These lead to the relationship for a metal $I = nAev$, where e is the electronic charge.

In a metal, a typical drift velocity is of the order of 1 mm s^{-1}, but in a semiconductor of similar dimensions carrying a similar current it is likely to be higher than this because the concentration of charge carriers is much less.

The charge carried by one electron, $e = -1.60 \times 10^{-19}$ C. 1 C of charge is equivalent to that carried by 6.25×10^{18} electrons.

A similar expression applies to non-metallic conductors. As the charge on each ion can vary between non-metals, the expression becomes $I = nAqv$, where q is the ionic charge.

Charge and energy transfer

Charge is not used up or changed into anything else as it moves round a circuit. The job it does is that of **transferring energy**. It gains energy from the source of electricity (mains, battery or power supply) and transfers this energy to the circuit components as it flows through them.

In the case of a filament lamp, the energy transfer from the charge is due to collisions of the electrons moving through the filament. With a motor or other electromagnetic device the charge loses energy as it has to work against repulsive forces in moving through the armature.

How much energy?

> The charge carriers gain energy as they are accelerated by the source; they lose energy in collisions.

Energy transfer is a continuous process as the charge flows round a circuit. There are two separate physical processes:

* energy transfer **to** the charge from the source
* energy transfer **from** the charge to the components

In each case, rather than measure the energy transfer for a single charge carrier, the transfer to and from **each coulomb** of charge is measured.

Electromotive force (e.m.f.) measures the energy transfer from the source.

A 6 V battery transfers 6 J of energy to each coulomb of charge that it moves around a circuit.

> Electromotive force is a mis-named quantity. It describes energy transfer rather than force.

> **KEY POINT**
>
> Electromotive force, symbol E, is defined as being the **energy transfer from the source in driving unit charge round a complete circuit**. It is measured in volts (V), where 1 volt = 1 joule/coulomb (1 J C^{-1}).

Electromotive force measures the work done **on** the charge **by** the source. As the charge flows in a circuit, a small amount of work is done in the connecting wires, transferring energy to heat, but most work is done in passing through the components.

Potential difference (p.d.) measures the work done by the charge in the circuit.

> **KEY POINT**
>
> The potential difference between two points, symbol V, is defined as being the **energy transfer per unit charge from the charge to the circuit**. Like e.m.f., it is measured in V.

* Electromotive force and potential difference are both defined in terms of the work done per unit charge, the difference being whether the work is done **on** the charge or **by** the charge.
* The definitions can be written as $E = W/q$ and $V = W/q$.
* The unit of e.m.f. and p.d. is the volt, where *one volt is the potential difference between two points if 1 J of energy is transferred when 1 C of charge passes between the points.*

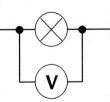

> The same instrument – a voltmeter – is used to measure both e.m.f. and p.d.

This voltmeter is measuring the e.m.f. of the cell; the energy transfer to each coulomb of charge

This voltmeter is measuring the p.d. across the lamp; the energy transfer from each coulomb of charge

Electrical power

AQA A	M3	NICCEA	M1
AQA B	M1	OCR A	M2
EDEXCEL A	M2	OCR B	M1
EDEXCEL B	M1	WJEC	M2

Unless stated otherwise, the term 'rate of' refers to 'rate of change with time', so power is the work done or energy transfer per second.

Power is the *rate of working or energy transfer*.

> **KEY POINT**
>
> The relationship between electrical power, current and potential difference is
> *power = current × potential difference* or *P=IV*.
> Power is measured in watts (W), where 1 W = 1 J s⁻¹.

This relationship can be combined with the resistance equation to give:

- $P = I^2R$
- $P = V^2/R$

These three equivalent relationships can all be used to calculate the value of either a constant or a varying power.

A battery-operated lamp transfers energy at a constant rate as there is no change in the current, potential difference or brightness of the filament. The power of a mains-operated lamp is constantly changing as all three of these variables change in step with each other. The graphs compare the power of a battery-operated lamp with that of a mains lamp.

The smaller the value of Δt, the better the approximation that $E = IV\Delta t$ is to the area between the graph line and the time axis.

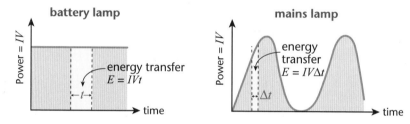

The SI unit of energy is the joule. For domestic purposes, energy is measured in kilowatt-hours (kW h), where 1 kW h is the energy transfer by a 1 kW appliance in 1 hour.

The total energy transfer in time Δt where the power is constant is equal to:

> **KEY POINT**
>
> $$\Delta E = power \times time = IV\Delta t.$$

For a varying power the energy transfer is represented by the area between the curve and the time axis of a graph of power against time. The shorter the value of Δt, the better the approximation of the above expression to this area.

Progress check

To answer i and ii, use the definitions of current and potential difference.

1 Charge flows through a lamp filament at the rate of 15 C in 60 s. In the same time, 1500 J of energy is transferred to the filament.
   a Calculate:
      i the current in the filament
      ii the potential difference across the filament
      iii the power of the lamp
      iv the resistance of the filament.
   b Explain how the power of the filament is affected when the current is doubled.

2 When the current in a cell is 1.5 A it supplies energy at the rate of 18 W. Calculate the e.m.f. of the cell and the total resistance of the circuit.

3 A silicon diode has a cross-sectional area of 4.0 mm² (4.0 × 10⁻⁶ m²). The concentration of charge carriers is 2.5 × 10²⁷ m⁻³ and each charge has a charge of 1.60 × 10⁻¹⁹ C. Calculate the drift velocity when the current in the diode is 1.5 A.

3 9.4×10^{-4} m s⁻¹
2 12 V and 8 Ω.
1 a i 0.25 A ii 100 V iii 25 W iv 400 Ω b The power becomes four times as great, since $P \propto I^2$.

2.3 Resistance in circuits

After studying this section you should be able to:

- calculate the effective resistance of a number of resistors connected in series or parallel
- understand that a source of electricity may contribute to the resistance of a circuit and the effects of this
- apply the principles of conservation of charge and conservation of energy to circuit calculations
- describe how the resistance of a metallic conductor, a thermistor and a light-dependent resistor change with environmental conditions

LEARNING SUMMARY

Series and parallel

AQA A	M3	NICCEA	M1
AQA B	M1	OCR A	M2
EDEXCEL A	M2	OCR B	M1
EDEXCEL B	M1	WJEC	M2

Resistors can be combined in series, parallel or a combination of the two. There are simple formulas for calculating the effective resistance of two or more resistors connected together.

Series

The effective resistance of a number of resistors in series is always greater than the largest value resistor in the combination.

Adding another resistor in series in a circuit always decreases the current. This means that the effective resistance in the circuit has increased.

> **KEY POINT**
>
> The formula for calculating the effective resistance of a number of resistors in series is:
> $$R = R_1 + R_2 + R_3.$$

This formula can be used for any number of resistors connected in series by adding on extra terms.

To find the effective resistance of a 4.7 Ω, a 6.8 Ω and a 10 Ω resistor connected in parallel
$1/R = 1/4.7 + 1/6.8 + 1/10$
$= 0.460$
$R = 1/0.460 = 2.2\ \Omega$
A common error is to forget to carry out the final stage of the calculation, in this case giving an answer of:
$R = 0.460\ \Omega$

Parallel

When an additional resistor is added in parallel to a circuit, the current always increases. The additional resistor opens up another current path, without affecting the current in any of the existing paths. This results in less resistance in the circuit.

The effective resistance of a number of resistors in parallel is always smaller than the smallest value resistor in the combination.

> **KEY POINT**
>
> The formula for calculating the effective resistance of a number of resistors in parallel is:
> $$1/R = 1/R_1 + 1/R_2 + 1/R_3.$$

As with the formula for resistors in series, any number of terms can be added to this expression.

Internal resistance

AQA A	M3	NICCEA	M1
AQA B	M1	OCR A	M2
EDEXCEL A	M2	OCR B	M1
EDEXCEL B	M1	WJEC	M2

All parts of a circuit have resistance and energy is needed to move the charge carriers through them. A cell of e.m.f. 1.5 V, for example, does 1.5 J of work on each coulomb of charge that completes a circuit.
This work is done on moving the charge through:

- the connecting wires
- the circuit components
- the cell itself, due to the cell's own internal resistance, symbol r.

The internal resistance of a cell or power supply acts as an extra resistor in series with the rest of the circuit.

The greater the current in the cell, the more work is done against the internal resistance, so less can be done in the external circuit.

How does internal resistance affect the circuit?

For example, a cell of e.m.f (E) 1.5 V and internal resistance (r) 0.2 Ω causes a current (I) of 0.5 A in an external resistor (R). The work done per coulomb of charge against the internal resistance $= I \times r = 0.5$ A $\times$ 0.2 Ω $= 0.1$ V. This leaves 1.4 V for the external circuit.

> The greater the current in the cell, the greater the reduction in potential difference.

- *The effect of internal resistance is to reduce the potential difference across the external circuit.*

When the only component connected to a cell is a high-resistance voltmeter such as a digital voltmeter, the reading is the cell's e.m.f. This is because when it is connected to the voltmeter alone, virtually no current passes in the cell so no work is being done against internal resistance. However, when the cell is connected to a circuit, a voltmeter placed across its terminals reads less than the e.m.f.

> A typical digital voltmeter has a resistance of 10 MΩ (1.0×10^7 Ω), so the current passing in it is very small, causing negligible reduction in the p.d. across the terminals.

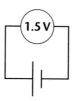

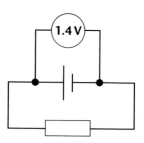

The e.m.f. of the cell is 1.5 V. The cell supplies 1.5 J of energy to move each coulomb of charge around a complete circuit.

0.1 J of energy is needed to move each coulomb of charge through the cell, leaving 1.4 J to move the charge round the rest of the circuit.

The difference between the e.m.f. and the 'terminal p.d.' is the work done within the cell against the internal resistance.

> **KEY POINT**
>
> The relationship between e.m.f. and terminal p.d. is:
> e.m.f. = terminal p.d. + p.d. across internal resistance
> $$E = V + Ir$$

Conservation rules

AQA A	M3	NICCEA	M1
AQA B	M1	OCR A	M2
EDEXCEL A	M2	WJEC	M2
EDEXCEL B	M1		

> Mass, charge and energy are fundamental conserved quantities. They cannot be created or destroyed.

In a circuit, the flow of **charge** transfers **energy** from the source to the components. Neither of these physical quantities becomes used up in the process. Both are conserved. Kirchoff's laws are re-statements of these fundamental laws in terms that apply to electric circuits.

> **KEY POINT**
>
> The first law states that the total current that enters a junction is equal to the total current that leaves the junction.

This is illustrated in the diagram and is simply a statement that charge is conserved at the junction.

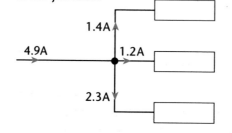

> The **second law** is about conservation of energy. It states that **around any closed loop (i.e. complete series path), the total e.m.f. is equal to the sum of the potential differences, $E = \Sigma IR$.**

Remember that e.m.f. represents energy transfer **to** the charge and p.d. represents energy transfer **from** the charge, so this law is stating that when a quantity of charge makes a complete circuit, the energy transfers each way are balanced. The second law can be applied to a series circuit or any series path within a parallel circuit.

> There are three closed loops in this parallel circuit. Can you identify the other two?

Resistance and the environment

AQA B	M1	NICCEA	M1
EDEXCEL A	M1	OCR A	M2
EDEXCEL B	M1	OCR B	M1
		WJEC	M2

Any change in environmental conditions, for example moisture level, temperature, pressure, stress and illumination can cause a change in resistance of a suitable sensor. The consequent change in current or potential difference can be used to monitor the forces in a structure or switch control circuits that perform tasks, such as the automatic watering of greenhouse crops and the maintaining of a constant temperature in an incubator.

The current in a **metallic conductor** is due to a flow of electrons, and resistance to the current is caused by anything that impedes that flow. Collisions between the free electrons and the metal structure become more frequent at higher temperatures, so the resistance increases. The change in resistance with temperature for a metallic conductor is small and approximately linear.

> Because of their large change in resistance with temperature, thermistors are useful in applications such as heat-sensitive cameras, which need to be able to detect small changes in temperature.

Much bigger changes in resistance occur when the temperature of a **thermistor** changes. Thermistors are made out of semi-conducting material; as the temperature rises the number of charge-carriers increases and the material becomes a better conductor. The effect of this swamps any increase in resistance due to increased collisions, and the resistance falls sharply with increasing temperature.

> These graphs show the variation in resistance with environmental conditions for a metallic conductor, a thermistor and a light dependent resistor.

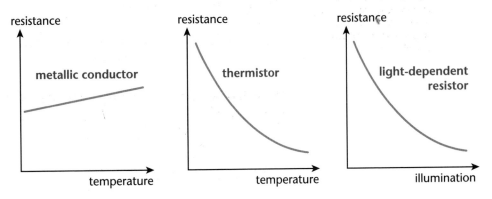

A **light-dependent resistor** (LDR) also shows large changes in resistance when the light level changes. Typically the resistance can vary from millions of ohms in total darkness to a few hundred ohms in daylight.

In a potential-divider circuit the voltage is shared in proportion to the resistance. If the resistance of one component increases, so does the voltage across it.

Thermistors and LDRs are widely used in control circuits, to switch on heaters and lamps for example. They are normally used in a potential-divider arrangement with a fixed resistor. In this arrangement the supply voltage is **shared** between the sensor (thermistor or LDR) and the fixed or variable resistor. If the resistance of a sensor changes, so does the potential difference across it; this can be used to switch an electronic circuit that controls other devices.

Progress check

In **b**, work out the resistance of the parallel combination first.

1 Calculate the effective resistance of each of the following combinations of resistors.

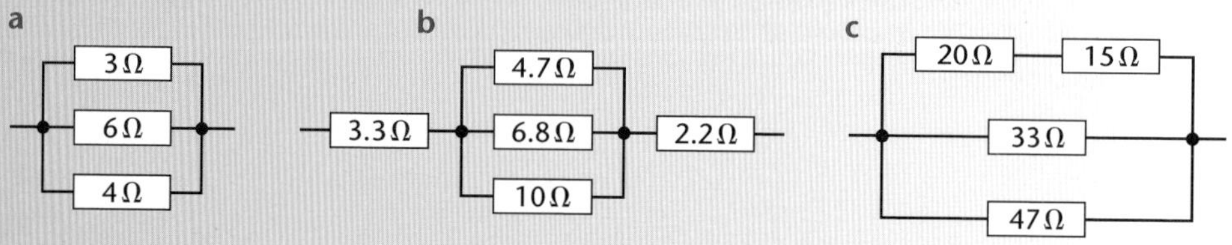

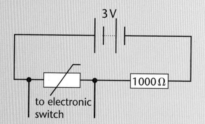

a
3 Ω
6 Ω
4 Ω
b
3.3 Ω
4.7 Ω
6.8 Ω
10 Ω
2.2 Ω
c
20 Ω
15 Ω
33 Ω
47 Ω

Remember that the p.d. across the terminals is equal to that across the external resistor.

2 A cell has an e.m.f. of 1.6 V and internal resistance 0.5 Ω.
Calculate the current in the cell and the p.d. across the terminals when it is connected to:

a a 0.5 Ω resistor
b a 3.5 Ω resistor.

3 A battery of e.m.f 4.5 V and internal resistance 0.6 Ω is connected to a 3.3 Ω resistor in series with a 5.1 Ω resistor.
Calculate:

a the current in the circuit
b the p.d. across each resistor
c the p.d. across the terminals of the cell.

4 The diagram shows a thermistor used in a potential divider as part of a temperature-controlled switch. The internal resistance of the power supply is small and can be neglected.

3 V
1000 Ω
to electronic switch

The electronic switch turns on a heater when the p.d. across the thermistor rises to 0.6 V.

The quickest way to answer **a** is to use ratios.

a Calculate the p.d. across the thermistor when its resistance is 200 Ω.
b Explain how the heater becomes switched off.
c Calculate the resistance of the thermistor when the heater switches off.

1 **a** 1.33 Ω **b** 7.67 Ω **c** 12.5 Ω
2 **a** 1.6 A; 0.8 V **b** 0.4 A; 1.4 V
3 **a** 0.50 A **b** 1.65 V and 2.55 V **c** 4.2 V
4 **a** 0.5 V **b** The temperature of the thermistor rises, causing its resistance and the p.d. across it to fall.
c 250 Ω.

2.4 Alternating current and the oscilloscope

After studying this section you should be able to:

- describe how to use an oscilloscope as a voltmeter and to measure time intervals and frequencies
- explain the meaning of root mean square, peak and peak-to-peak values of a sinusoidal waveform

LEARNING SUMMARY

Alternating and direct current

AQA A ▸ M3
NICCEA ▸ A2
WJEC ▸ A2

The current from a battery is a **direct current** (d.c.), while that from the mains is an **alternating current** (a.c.).

- An alternating current changes direction.
- A direct current may vary in value, but maintains the same direction.

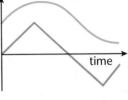

A direct current and an alternating current. Which is which?

Controlling an oscilloscope

AQA A ▸ M3
NICCEA ▸ A2

An oscilloscope is a high-resistance voltmeter that can display how a voltage changes over a period of time. An electron beam strikes the screen and causes it to fluoresce, making a bright dot on the screen.

The oscilloscope has two key controls:

- the **sensitivity** or **gain** controls the vertical deflection of the dot
- the **time base** controls the speed of horizontal movement of the dot.

The **sensitivity** control is marked in 'volts/cm' or 'volts/div', this is the voltage required to cause a 1 cm vertical deflection.

Because of its high resistance, an oscilloscope can only be used as a voltmeter to study the waveform of a voltage. It cannot be used in place of an ammeter to study the waveform of a current.

The **time base** control is marked in 'time/cm' or 'time/div', the setting being the time it takes for the dot to move horizontally one centimetre. The shorter this time interval, the faster the dot moves across the screen. With all but the slowest settings, the dot moves so fast that it appears to be a horizontal line.

time base control

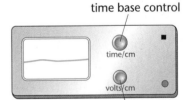

sensitivity control

These diagrams compare the traces obtained for a 3.0 V direct voltage input with different sensitivity and time base settings.

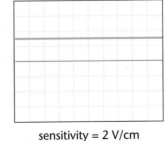

sensitivity = 2 V/cm
time base on

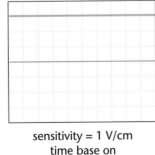

sensitivity = 1 V/cm
time base on

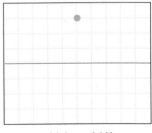

sensitivity = 1 V/cm
time base off

Measuring time and frequency

AQA A   M3
NICCEA   A2
WJEC   A2

The time base setting on the oscilloscope can be used to measure short time intervals and the frequency of an alternating voltage.
- A time interval is measured by multiplying the appropriate horizontal distance on the screen by the time base setting.
- A frequency is measured by calculating (1/time) for one complete cycle of a wave.

> Remember to convert times in ms or μs into s before calculating a frequency.

In the example shown in the diagram, the time base is set to 1 ms cm⁻¹, or 1.0×10^{-3} s cm⁻¹. One complete cycle of the wave occupies a distance of 4.0 cm on the screen, so the time taken for one cycle is 4.0×10^{-3} s.
Since *frequency = 1/time period* or $f = 1/t$, the frequency of the wave is
$1 \div 4.0 \times 10^{-3}$ s = 250 Hz.

Measuring alternating voltage

AQA A   M3
NICCEA   A2
WJEC   A2

Alternating voltage from the mains varies in the form of a sine wave. This means that it not only changes direction, it is also continually changing its value. Two measurements that can be made from an oscilloscope trace are:
- the **peak value**, V_0, the maximum or minimum value of the voltage
- the **peak-to-peak value**, the difference between the maximum and minimum values.

> Normally the peak-to-peak value of an alternating voltage is twice the peak value.

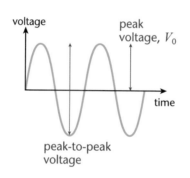

The peak value of an alternating current or voltage can be a misleading measurement, since for most of the time the current or voltage has a lower value than this. To make a fair comparison with a direct current or voltage, the **root mean square** or **r.m.s** value is quoted. The r.m.s value of an alternating current or voltage has the same heating effect as a direct current or voltage of the same value. This means that a lamp lights with the same brightness when connected to 12 V d.c. and 12 V r.m.s a.c.

The relationship between the r.m.s value of an alternating current or voltage and the peak value is:

> Mains voltage is described as 240 V. This is an r.m.s value, the peak value being $240 \times \sqrt{2} = 339$ V.

$$I_{r.m.s} = \frac{I_0}{\sqrt{2}} \qquad V_{r.m.s} = \frac{V_0}{\sqrt{2}}$$

KEY POINT

This relationship is only valid for alternating currents and voltages that are sinusoidal, i.e. that follow the pattern of a sine wave.

Progress check

The diagram shows an oscilloscope display of an alternating voltage.

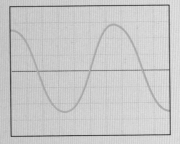

The sensitivity is set to 10 V/cm and the time base to 5 ms/cm. Calculate:

i the peak value of the voltage

ii the r.m.s. value of the voltage

iii the frequency of the voltage.

i 25 V   ii 17.7 V   iii 33.3 Hz

2.5 Electromagnetism

After studying this section you should be able to:

- sketch the magnetic field patterns due to a current in a straight wire, a circular coil and a solenoid
- use Fleming's left-hand rule to predict the direction of the force on a current-carrying conductor placed in a magnetic field
- define magnetic field strength (flux density) and the tesla
- recall and use $F = BIl$

Magnetic field patterns

AQA A	A2	NICCEA	A2
AQA B	A2	OCR A	M2
EDEXCEL A	A2	OCR B	A2
EDEXCEL B	A2	WJEC	A2

At AS/A level you study three types of field: magnetic fields, gravitational fields and electric fields. Take particular note about the similarities and differences between them.

A field is a name given to a region where forces are exerted on objects with a certain property. A gravitational field describes a region where there are forces on objects that have mass. An electric field affects objects that have charge. A magnetic field affects:

- permanent magnets
- magnetic materials
- electric currents.

You should be familiar with magnetic field patterns drawn around permanent magnets. Although the force at any point is in a straight line, the force lines joined together to make a series of curved lines. Arrows on the magnetic field patterns show the direction of the force on the north-seeking pole of a magnet.

Similarly, every electric current has its own magnetic field. The diagrams show the field patterns due to the currents passing in:

- a long straight wire
- a circular coil
- a solenoid (a long coil).

Notice that the magnetic field in the centre of the solenoid is uniform, while on the outside it resembles that of a bar magnet.

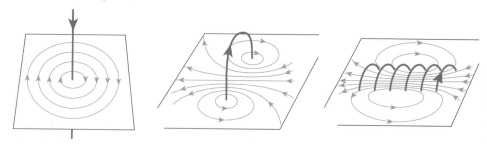

Magnetic forces

AQA A	A2	NICCEA	A2
AQA B	A2	OCR A	M2
EDEXCEL A	A2	OCR B	A2
EDEXCEL B	A2	WJEC	A2

Electric motors make use of the force that acts on a current-carrying conductor placed at right angles to a magnetic field. Provided that it is not parallel to the field (in which case there is no force), any electric current in a magnetic field experiences a force.

- The maximum force occurs when the current and magnetic field are at right angles to each other.
- The direction of the force is at right angles to both the field and the current.
- Fleming's left-hand rule shows the relationship between the directions of the current, magnetic field and force.

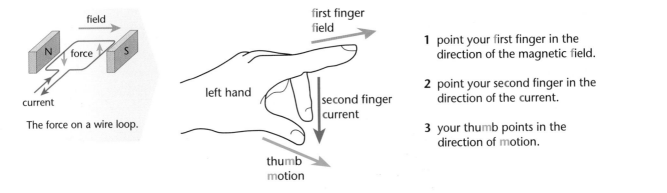

field / force / current

The force on a wire loop.

first finger
field

left hand

second finger
current

thumb
motion

1 point your first finger in the direction of the magnetic field.

2 point your second finger in the direction of the current.

3 your thumb points in the direction of motion.

The size of the force

AQA A	A2	NICCEA	A2
AQA B	A2	OCR A	M2
EDEXCEL A	A2	OCR B	A2
EDEXCEL B	A2	WJEC	A2

The size of the force that acts on a current-carrying conductor in a magnetic field depends on:

- the direction of the current relative to the magnetic field
- the size of the current
- the length of conductor in the magnetic field
- the strength of the magnetic field.

> **KEY POINT**
> These factors are used to define the strength of a magnetic field as:
> **the force per unit (current × length) perpendicular to the field.**
> Magnetic field strength is also known as *flux density*. Its symbol is B and the unit is the tesla (T).

With the current and magnetic field directions shown in the diagram, the force is into the paper and the definition of magnetic field strength can be written as:

$B = F/Il$ or $F = BIl$.

> You must always specify the directions of the quantities when using this formula to define magnetic field strength.

l

B

I

> **KEY POINT**
> This equation is also used to define the size of the tesla as:
> **the magnetic field strength that exerts a force of 1 N on a (current × length) of 1 A m placed at right angles to the field.**

The force between two parallel conductors like those in Q2 below is now used as the basis for the definition of the unit of current, the ampère.

Progress check

1 a Draw a clear, labelled diagram that shows how magnetic field strength is defined in terms of the force on a current-carrying conductor. Your diagram should show the directions of any vector quantities.

   b The magnetic field strength in the centre of a solenoid is 2.0×10^{-4} T. Calculate the size of the force on a wire of length 0.15 m carrying a current of 2.5 A placed:

   i along the axis of the solenoid

   ii perpendicular to the axis of the solenoid.

2 Two parallel wires carry currents in opposite directions. Draw a diagram showing the magnetic field due to each wire and the direction of the force on each wire.

2 The diagram should show that each wire is repelled from the other wire.

1 a see text b i 0 ii 7.5×10^{-5} N

Sample question and model answer

(a) Show that the unit of resistivity is the Ω m. [1]

Since $\rho = AR/l$, the units of ρ must equal those of AR/l.
Units of $AR/l = m^2 \times \Omega \div m = \Omega\ m$.

This technique relies on the fact that a relationship between physical quantities must be homogenous, i.e. the units on each side are the same.

(b) A cable consists of seven straight strands of copper wire each of diameter 1.35 mm as shown in the diagram. [4]

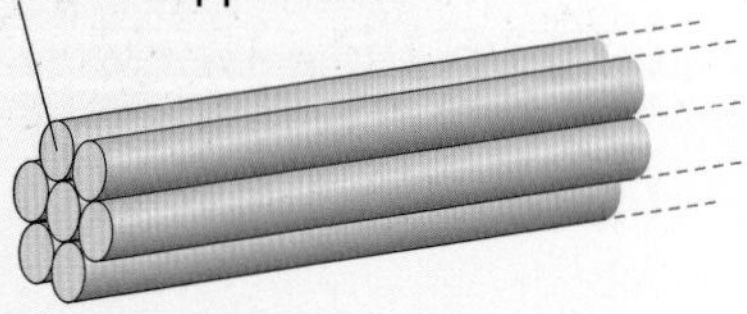

strand of copper wire

Calculate:

(i) the cross-sectional area of **one strand** of copper wire.

It is important here to convert the diameter from mm to m in order to maintain a consistent set of units. Many AS candidates, faced with this type of calculation, work out the answer in mm² and then apply a wrong conversion factor to change it to m².

Cross-sectional area $= \frac{1}{4}\pi d^2$
$= \frac{1}{4} \times \pi \times (1.35 \times 10^{-3})^2 = 1.43 \times 10^{-6}\ m^2$ 1 mark

(ii) the resistance of a 100 m length of the cable, given that the resistivity of copper is 1.6×10^{-8} Ω m.

The resistance of one strand,
$R = \rho l/A$ **(1 mark)**
$= 1.6 \times 10^{-8}\ \Omega\ m \times 100\ m \div 1.43 \times 10^{-6}\ m^2 = 1.12\ \Omega$ 1 mark
The resistance of 100 m of the cable is $\frac{1}{7}$ of this, i.e. 0.16 Ω.   1 mark

The emphasis here is on reading the question thoroughly. A common error is to calculate the resistance of one strand only, ignoring the fact that the resistance of the cable is emphasised in the question.

(c) (i) If the cable in part (b) carries a current of 20 A, what is the potential difference between the ends of the cable? [2]

$V = IR = 20\ A \times 0.16\ \Omega = 3.2\ V.$ 1 mark

(ii) If a single strand of the copper wire in part (b) carried a current of 20 A, what would be the potential difference between its ends?

$20\ A \times 1.12\ \Omega = 22.4\ V.$ 1 mark

The calculations in **c** are straightforward applications of the resistance equation. Do not worry if your answers to **b** were wrong. You still gain full marks in **c** for correct working with the wrong resistance values.

(d) State **one** advantage of using a stranded rather than a solid core cable with copper of the same total cross-sectional area. [1]

A stranded cable is more flexible than a solid one and is less prone to snap due to repeated bending.

NEAB Mechanics and Electricity (PH01), June 1999 Q 2

The fixed wiring that connects the sockets in a house back to the consumer unit is done with solid cables as these are not subjected to repeated flexing. Connections between an appliance and a mains socket use stranded cables.

Practice examination questions

1 (a) Write down the formula that relates the resistivity of a material to the resistance of a particular sample. [1]

(b) Explain the difference between *resistance* and *resistivity*. [2]

(c) The diagram shows a cuboid made from carbon.

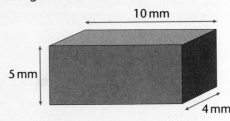

The resistivity of carbon is 3.00×10^{-5} Ω m.
Calculate the resistance of the cuboid, measured between the faces that are 10 mm apart. [3]

2 In the circuit shown in the diagram, the internal resistance of the cell can be neglected.

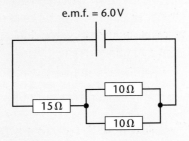

Calculate:
(a) the effective resistance of the two resistors in parallel [2]

(b) the current in the 15 Ω resistor [3]

(c) the current in each 10 Ω resistor [2]

(d) the potential difference across each 10 Ω resistor. [2]

3 (a) (i) Draw a sketch graph to show how the resistance of a thermistor varies with increasing temperature. [2]

(ii) State TWO ways in which this graph differs from a graph that shows how the resistance of a metallic conductor varies with increasing temperature. [2]

(iii) Explain why Ohm's law does not apply to either graph. [2]

(b) The diagram shows a thermistor in a series circuit with a 100 Ω resistor.

Practice examination questions (continued)

The table gives some data about the resistance of the thermistor.

temperature/°C	thermistor resistance/Ω
5	500
20	300
40	75

(i)   Calculate the potential difference across the thermistor at a temperature of 5°C.   [3]

(ii) Explain how the potential difference across the thermistor changes as the temperature rises.   [2]

(iii) The potential difference across the resistor is used to switch off a heater when the temperature reaches 40°C. Calculate the potential difference at which the switch operates.   [3]

(iv) The circuit is adapted to switch off the heater when the temperature reaches 20°C. Calculate the value of the fixed resistor required.   [3]

4 A cell has an e.m.f. of 1.60 V. When connected to a digital voltmeter of resistance 10 MΩ (1.0×10^7 Ω), the reading on the voltmeter is 1.60 V. When the cell is connected to a moving coil voltmeter of resistance 1 kΩ (1.0×10^3 Ω) the reading on the voltmeter is 1.55 V.

(a) Explain why there is a difference in the voltmeter readings.   [3]

(b) Calculate the internal resistance of the cell.   [3]

5 A 12 V car battery is used to light four 6 W parking lamps connected in parallel.

(a) Calculate the current in the battery.   [1]

(b) How much charge flows through the battery in one minute?   [3]

(c) Calculate the effective resistance of the four lamps.   [3]

6 (a) A 4.7 Ω resistor has a power rating of 10 W.

(i)   Calculate the maximum current that should pass in the resistor.   [3]

(ii) Calculate the maximum potential difference across the resistor.   [2]

(iii) The resistor is manufactured from constantan wire of cross-section 2.0×10^{-7} m² and resistivity 5.0×10^{-7} Ω m. Calculate the length of wire used to make the resistor.   [3]

(b) Low-power resistors are often made from carbon. The resistivity of carbon is 3.0×10^{-5} Ω m. What is the advantage of making resistors from carbon?   [2]

Practice examination questions *(continued)*

7   The diagrams show two circuits that can be used to investigate the current–voltage characteristics of component X.

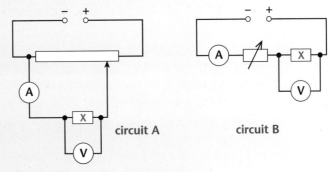

circuit A   circuit B

(a) What name is given to circuit A?   [1]

(b) Describe what circuit A is able to do.   [1]

(c) What is the advantage of using circuit A rather than circuit B to investigate the characteristics of a diode?   [2]

(d) The table shows data obtained by varying the voltage across a diode.

voltage/V	0.65	0.71	0.76	0.78	0.80
current/A	0.06	0.20	0.44	0.68	1.00
resistance/Ω					

(i)   Complete the table.   [2]

(ii) Draw a graph of resistance against voltage.   [4]

(iii) Use the graph to estimate the voltage at which the resistance of the diode is 1 Ω.   [1]

(iv) Suggest why the data only span a narrow voltage range.   [1]

8   In a simple lighting circuit, a 60 W lamp is connected to a 12 V battery using copper cable of cross section 1.25 mm^2 (1.25×10^{-6} m^2). The lamp filament is made of tungsten of cross section 3.0×10^{-4} mm^2 (3.0×10^{-10} m^2). The number of free electrons per unit volume for copper is 8.0×10^{28} m^{-3} and that for tungsten is 3.4×10^{28} m^{-3}. The electronic charge $e = -1.60 \times 10^{-19}$ C.

(a) Calculate the drift velocity of the electrons in the copper cable.   [3]

(b) Give TWO reasons why the drift velocity of the electrons in the tungsten is much greater than that of electrons in the copper.   [2]

(c) Suggest why the tungsten becomes heated while the copper remains cool.   [2]

9   A cell of e.m.f. 3.0 V and internal resistance 0.5 Ω is connected to a 3.0 Ω resistor placed in series with a parallel combination of a 6.0 Ω and a 12.0 Ω resistor.

Calculate:

(a) the current in the circuit   [3]

(b) the p.d. across the terminals of the cell.   [2]

Chapter 3

Particle physics

The following topics are covered in this chapter:

- *A model gas*
- *Internal energy*
- *Radioactive decay*

- *Forces and particles*
- *Stability and the nucleus*

3.1 A model gas

After studying this section you should be able to:

- *understand the concept of absolute zero of temperature and be able to convert temperatures between the Celsius and Kelvin scales*
- *recall and use the ideal gas equation of state pV = nRT*
- *explain the relationship between the kelvin temperature and the mean kinetic energy of the particles of an ideal gas*

LEARNING SUMMARY

Gas pressure

AQA A	M2	NICCEA	A2
AQA B	A2	OCR A	A2
EDEXCEL A	M2	OCR B	A2
EDEXCEL B	A2	WJEC	A2

As you will see later in this section, the average speed of the particles is different for different gases.

Brownian motion was first observed by Robert Brown while studying pollen grains suspended in water. The explanation of Brownian motion is due to Einstein.

The kinetic model pictures a gas as being made up of large numbers of individual particles in constant motion. At normal pressures and temperatures the particles are widely-spaced compared to their size. The motion of an individual particle is:

- **rapid** – a typical average speed at room temperature is 500 m s⁻¹
- **random** in both speed and direction – these are constantly changing due to the effects of collisions.

Gases exert **pressure** on the walls of their container and any other objects that they are in contact with. This pressure acts in all directions and is due to the forces as particles collide and rebound.

Evidence for this movement of gas particles comes from **Brownian motion**; the random, lurching movement of comparatively massive particles such as smoke specks when suspended in air. This movement can only be attributed to the bombardment by much smaller air particles which are too small to be seen by an optical microscope and must therefore be moving very rapidly.

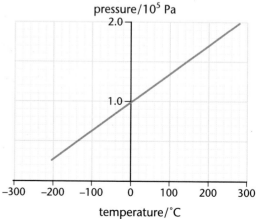

Gas particles are widely spaced. Their movement is random in both speed and direction.

Experiments show that increasing the temperature of a gas also increases the pressure that it exerts. At higher temperatures the average speed of the particles increases. This affects the pressure in two ways:

- it increases the average force of the collisions
- collisions are more frequent.

The graph shows how the pressure of a gas depends on its Celsius temperature when the volume of the gas is kept constant. Although the graph is a straight line, it does not show that pressure is directly proportional to Celsius temperature since the line does not go through the origin.

This is not surprising since 0°C does not represent 'no temperature'; it is merely a reference point that is convenient to use because it is easily reproduced.

The graph does however give an indication of whereabouts 'no temperature' is on the Celsius scale. If temperature is a measure of particle movement, then the minimum possible temperature corresponds to the minimum amount of particle movement. By extrapolating the graph back to the corresponding pressure, zero, it gives a figure for the minimum temperature at –273°C.

Knowledge of the zero of temperature allows it to be measured on an absolute scale. On an absolute scale of measurement:

- the zero is the minimum amount of a quantity – it is not possible for a smaller measurement to exist
- a measurement of 40 units is twice that of 20 units.

> The extrapolation assumes that the substance remains a gas as it is cooled. Real gases would reach their boiling point and liquefy before the minimum temperature was reached.

> Note that the kelvin is not called a degree and does not have a ° in the unit.

> **KEY POINT**
>
> The absolute scale of temperature is called the Kelvin scale; its unit is the kelvin (K) and its relationship to the Celsius scale is:
> $$T/K = \theta/°C + 273$$

The diagram shows the relationship between Celsius and kelvin temperatures.

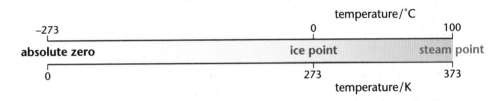

When Celsius temperatures are converted to kelvins, the experimental results show that:

> The value of the constant depends on the amount of gas and its volume.

> **KEY POINT**
>
> For a fixed mass of gas at constant volume:
> pressure ∝ absolute temperature or p/T = constant

> Remember: the relationship is only valid when temperature is measured in kelvin.

The ideal gas equation

AQA A	M2	NICCEA	A2
AQA B	A2	OCR A	A2
EDEXCEL A	M2	OCR B	A2
EDEXCEL B	A2	WJEC	A2

When experimenting with a fixed mass of gas the three variables are pressure, volume and temperature. Fixing one of these enables the relationship between the other two to be established. In addition to the relationship between pressure and temperature above:

> The relationship $pV =$ constant is known as Boyle's law.

> **KEY POINT**
>
> For a fixed mass of gas at constant temperature, pressure is inversely proportional to volume:
> pressure ∝ 1/volume   or pV = constant
> For a fixed mass of gas at constant pressure, the volume is proportional to absolute temperature:
> volume ∝ temperature   or V/T = constant

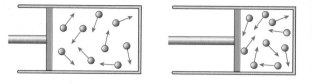

Halving the volume of a gas leads to twice the rate of collisions – doubling the pressure.
This is an example of Boyle's Law.

The concept of an ideal gas is useful because the behaviour of all gases is close to ideal provided that certain conditions about temperature and pressure are met.

These relationships are based on experiments and are valid for all gases provided that the gas is neither at a high pressure nor close to its boiling point. They lead to the concept of an **ideal gas** as one that obeys Boyle's law, i.e. a gas for which pV = constant.

Real gases do not obey Boyle's law when the particles are close enough together so that they occupy a significant proportion of the gas's volume and exert appreciable forces on each other.

For an ideal gas the three gas laws can be combined into the single equation:

pV/T = constant

1 mole of gas consists of the Avagadro constant, N_A, of particles. The value of the Avagadro constant is 6.02×10^{23} mol⁻¹.

In this case the value of the constant depends only on the amount of gas. For 1 mole of any gas under ideal conditions it has the value of 8.3 J mol⁻¹ K⁻¹. This constant is called the **molar gas constant** and has the symbol R. Doubling the number of particles by considering two moles of gas has the effect of doubling the value of the constant, so:

> The equation of state for an ideal gas is:
> $$pV = nRT$$
> where n is the number of moles of gas and R is the molar gas constant.
>
> **KEY POINT**

An algebraic model

If a gas is sealed in a container, then increasing its temperature results in an increase in the pressure of the gas. This can be attributed to an increase in the average speed of the particles. The kinetic model can be extended to establish the relationship between the temperature of a gas and the speed of the particles. First of all it is necessary to establish just what is meant by 'ideal gas behaviour'.

The kinetic theory of gases involves some basic assumptions about the particles of a gas:

- a gas consists of a large number of particles in a state of rapid, random motion
- gas pressure is a result of collisions between the particles and the container walls.

In addition, for an ideal gas it is assumed that:

- the volume occupied by the particles is negligible
- intermolecular forces are negligible
- all collisions are elastic, so there is no loss in kinetic energy.

With these assumptions, it can be shown that **the pressure of an ideal gas is proportional to the mean of the squares of the particle speeds**:

Note that working out $< c^2 >$ involves:
- listing all the individual particle speeds
- squaring them
- calculating the mean of the squares.

> $$pV = 1/3 \, Nm < c^2 >$$
> Where N = total number of particles
> m = mass of each particle
> $< c^2 >$ = mean of the squares of the particle speeds
>
> **KEY POINT**

Alternatively, since Nm/V = mass ÷ volume = density:

> $$p = 1/3 \, \rho < c^2 >$$
> Where ρ = density of the gas
>
> **KEY POINT**

If the mass and volume of an ideal gas are fixed, the gas pressure is proportional to its temperature, p/T = constant. The above results show that, as pressure is proportional to $< c^2 >$ it must be proportional to the mean kinetic energy of the particles.

These factors taken together give the relationship between kinetic energy and temperature:

> **KEY POINT**
>
> the mean kinetic energy of the particles of an ideal gas is proportional to the kelvin temperature
>
> $$\tfrac{1}{2}m< c^2 > \propto T$$

> This does not mean that hydrogen molecules move, on average, four times as fast as those of oxygen at a particular temperature. The root mean square speed, $\sqrt{< c^2 >}$, can be used as a typical speed but not the mean speed of the particles.

This means that doubling the kelvin temperature of a gas doubles the mean kinetic energy and the mean square speed of its particles. Remember that these relationships apply to all ideal gases, irrespective of the gas. So at a particular temperature the mean kinetic energy of the molecules in a sample of hydrogen is the same as those in a sample of oxygen. Since an oxygen molecule has sixteen times the mass of a hydrogen molecule, the mean square speed of the hydrogen molecules must be sixteen times as great as that of oxygen molecules.

The Boltzmann constant

AQA A	M2	NICCEA	A2
AQA B	A2	OCR B	A2
EDEXCEL B	A2	WJEC	A2

The algebraic model can be used to put a figure on the mean kinetic energy of the particles of an ideal gas at any temperature.

For one mole of gas the number of particles, N is equal to Avagadro's number, N_A. So for one mole, combining the equations $pV = RT$ and $pV = 1/3 N_A m < c^2 >$ gives:
$$\tfrac{1}{2}m < c^2 > = \tfrac{3}{2}RT/N_A \text{ or } N_A \times \tfrac{1}{2}m < c^2 > = \tfrac{3}{2}RT$$
This means that the total kinetic energy of one mole of an ideal gas is equal to $\tfrac{3}{2}RT$, since N_A is the number of particles and $\tfrac{1}{2}m < c^2 >$ is their mean kinetic energy.

The Boltzmann constant, k, relates the mean kinetic energy of the particles in an ideal gas directly to temperature:

> **KEY POINT**
>
> $$\tfrac{1}{2}m < c^2 > = \tfrac{3}{2}RT/N_A$$
> $$= 3/2\ kT$$
> Where $k = R/N_A$ is the Boltzmann constant and has the value 1.38×10^{-23} J K^{-1}

> To work out the mass of a molecule of a gas, divide the molar mass by the Avagadro constant.

So, for example, at a temperature of 20°C (293 K) the mean kinetic energy of the particles of an ideal gas is equal to $3/2 \times 1.38 \times 10^{-23}$ J K$^{-1} \times 293$ K $= 6.07 \times 10^{-21}$ J. This may seem a tiny amount of energy, but remember that it refers to a single gas particle which has a very small mass.

Progress check

Assume that $R = 8.3$ J mol^{-1} K^{-1}

1 2.0 mol of an ideal gas is sealed in a container of volume 5.5×10^{-2} m^3.
   a Calculate the pressure of the gas at a temperature of 18°C.
   b At what Celsius temperature would the pressure of the gas be doubled?

2 1.0 mol of oxygen has a mass of 32 g. Calculate the mean square speed of the molecules at 0°C and a pressure of 1.01×10^5 Pa.

3 A room measures 3.5 m × 3.2 m × 2.2 m. It contains gas at a pressure of 1.03×10^5 Pa and a temperature of 25°C. Calculate the number of moles of gas in the room.

3 1026
2 2.1×10^5 m^2 s^{-2}
b 309°C
1 a 8.8×10^4 Pa

3.2 Internal energy

After studying this section you should be able to:

- *explain the meaning of* specific heat capacity *and* specific latent heat *and use the appropriate relationships*
- *describe the principle of operation of a* heat engine *and a* heat pump
- *calculate the maximum efficiency of a heat engine*

LEARNING SUMMARY

Energy of individual particles

AQA A	M2	NICCEA	A2
AQA B	A2	OCR A	A2
EDEXCEL A	M2	OCR B	A2
EDEXCEL B	A2	WJEC	A2

All materials are made up of particles, and particles have energy. The particles of an ideal gas have only kinetic energy since forces between the particles are negligible. However, this cannot be true of a solid or a liquid or the particles would not stay in that phase, and the particles in real gases exert both attractive and repulsive forces on each other during collisions.

Imagine a collision between two particles in a real gas:

- as the particles approach, the attractive forces cause an increase in the speed and kinetic energy
- the particles then get closer and the speed decreases, being momentarily zero before they start to separate
- the process is then reversed.

> The speed of colliding particles is momentarily zero as their direction of travel is reversed.

There is an interchange between potential (stored) energy and kinetic energy during the collision, as shown in the diagram.

   pe → ke

   ke → pe

> As the separation between two particles decreases, the net force changes from attractive to repulsive. The equilibrium position is where these forces are equal in size, so the resultant force is zero.

particles speed up due to short-range attractive forces

particles slow down due to close-range repulsive forces

The particles of real gases, like those of solids and liquids, have both potential and kinetic energy. The distribution of energy between kinetic and potential is constantly changing and so is said to be random.

The total amount of energy of the particles in an object is known as its **internal energy**. At a constant temperature the internal energy of an object remains unchanged, but the contributions of individual particles to that energy change due to the transfer of energy during interactions between particles.

Being specific

AQA A	M2	NICCEA	A2
AQA B	A2	OCR A	A2
EDEXCEL A	M2	OCR B	A2
EDEXCEL B	M1	WJEC	A2

Changing the temperature of an object involves a change in its internal energy, the total potential and kinetic energy of the particles. One way of doing this is to place the object in thermal contact with a hotter or colder object, so that there is a net flow of energy from hot to cold (see diagram).

> Two objects are in thermal contact if they can exchange heat. Energy flows both ways between a warm and a cold object in thermal contact, but there is more energy flow from the warm object to the cold one than the other way.

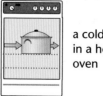

a cold dish in a hot oven

warm food placed in a refrigerator

The energy flow between objects at different temperatures

The energy transfer required to change the temperature of an object depends on:

- the temperature change
- the mass of the object
- the material the object is made from.

These are all taken into consideration in the concept of **specific heat capacity**. The term 'specific' means 'for each kilogram', so any physical measurement that is described as 'specific' refers to 'per kilogram of material'.

> When using this relationship the temperatures can be in either Celsius or kelvin since the temperature change is the same in each case.

> **KEY POINT**
>
> The specific heat capacity of a material, c, is defined as:
> The energy transfer required to change the temperature of 1 kg of the material by 1°C.
> $$\Delta E = mc\,\Delta\theta$$
> where c, specific heat capacity, is measured in J kg^{-1} °C^{-1} or J kg^{-1} K^{-1}

Changing phase

AQA A ▶ M2   EDEXCEL A ▶ M2
AQA B ▶ A2   OCR A ▶ A2
             OCR B ▶ A2

If you try to squash a solid or a liquid, the particles are pushed closer together and they repel each other. Stretching has the opposite effect; when the separation of the particles is increased the forces are attractive. At increased separations the particles have increased potential energy and this energy has to be supplied for any process that involves expansion to take place.

> The term 'phase' means whether the substance is a solid, liquid or gas.

When a substance changes **phase** there is a change in the potential energy of the particles. For most substances there is a small increase in potential energy when changing from solid to liquid and a much bigger change from liquid to gas.

> The increase in particle separation during a change of phase from solid to liquid is small, but that during a change of phase from liquid to gas is large.

solid   **liquid**   **gas**

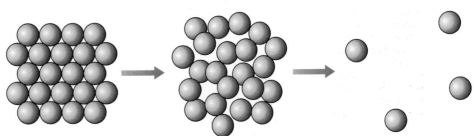

There is a small increase in potential energy of the particles when a substance changes from solid to liquid, and a much larger increase in changing from liquid to gas.

The energy absorbed or released during a change of phase is called **latent heat**. As with heat capacity, the term 'specific latent heat' refers to 1 kilogram of material.

> This definition applies to a change of phase where energy is removed from the substance, as well as a change where energy is supplied.

> **KEY POINT**
>
> The specific latent heat of a material, l, is defined as:
> The energy required to change the phase of 1 kg of the material without changing its temperature.
> $$E = ml$$

Note that there are two values of the specific latent heat for any material:

- the specific latent heat of fusion refers to a change of phase between solid and liquid
- the specific latent heat of vaporisation refers to a change of phase between liquid and gas

On each horizontal part of the curve the substance exists in two phases. Can you identify them?

The diagram shows a typical temperature–time graph as a solid is heated at a constant rate and passes through the three phases. Where the curve is horizontal it shows that energy is being absorbed with no change in temperature. This corresponds to a change of phase. Note that much more energy is absorbed during the change from liquid to gas than during the change from solid to liquid.

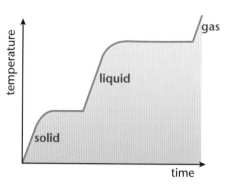

Heat and work

AQA B	A2	OCR A	A2
EDEXCEL A	M2	WJEC	A2

Placing an object in thermal contact with one that is hotter or colder, for example an immersion heater or the inside of a freezer, is one way of changing its temperature. Anyone who has ever pumped up a bicycle tyre is aware that compressing the air in the pump also causes it to become warmer.

Compressing a gas causes heating and expanding a gas causes cooling. Whether the gas is heated or cooled depends on whether work is done **on** it or **by** it.

Other everyday examples of work causing a change in temperature include:

- striking a match, where work done against friction forces results in heating of the match head
- refrigeration, where the rapid expansion of a liquid into a vapour causes cooling.

The **first law of thermodynamics** states the internal energy of a gas depends only on its state, i.e. the conditions of temperature, pressure and volume and the amount of gas, and not how it reached that state. Transferring energy as heat to a gas has the same effect as doing work on it – they both cause an increase in the internal energy, and therefore the temperature.

In other words, the first law states that work and heat are equivalent.

> **KEY POINT**
>
> The first law of thermodynamics can be written as:
> increase in internal energy of gas = heat transferred to gas + work done on gas
> $$\Delta U = Q + W$$
> Note that the values of ΔU, Q and W are negative if the internal energy decreases or heat is removed from the gas or work is done by the gas.

Heat engines

EDEXCEL A	M2

Internal combustion engines drive cars, buses and some trains. In these engines:

- the burning of fuel supplies heat to a gas, increasing its internal energy
- the gas does work as it pushes the pistons, decreasing the internal energy.

In an internal combustion engine, the pistons are driven by the hot gases produced when fuel burns. Energy that is not removed from the gases is wasted in the exhaust gases and passes to the atmosphere, which is at a lower temperature.

Rockets, jet engines and steam engines are all examples of heat engines.

This illustrates what is meant by a **heat engine**; a device that uses the energy flow from a **hot source** to a **cold sink** to do work. The principle of operation of a heat engine is shown in the diagram.

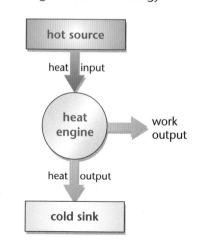

It is not possible to extract all the internal energy from the steam in a steam turbine. Can you explain why?

The steam turbines that drive the generators in power stations are also heat engines. Steam enters the turbines at high temperature and pressure and leaves at much lower temperature and pressure. Of the internal energy lost by the steam:

- some does work on the turbines
- some is lost to the surroundings.

The turbines are designed to extract as much energy from the steam as possible; the greater the proportion of the available energy extracted, the more efficient the process is.

If there are no energy losses, then the maximum efficiency of a heat engine is given by the relationship:

Refer to page 62 for the definition of efficiency.

> **KEY POINT**
>
> maximum efficiency = $(T_H - T_C) \div T_H$
> where T_H is the temperature of the hot source and T_C is the temperature of the cold sink. Both temperatures are measured in kelvin.

For example, if steam enters a turbine at 250°C (523 K) and leaves at 120°C (393 K) then the maximum efficiency is (523 K – 393 K) ÷ 523 K = 0.25. In practice, real heat engines do not operate at their maximum efficiency since there are always other energy losses.

Heat pumps

EDEXCEL A   M2

Some ovens and convector heaters use fans to speed up the rate of energy transfer.

When two objects at different temperatures are placed in thermal contact, the net energy flow is always from hot to cold. No other energy source is needed to drive this flow, although sometimes fans are used to speed up the natural process. In refrigeration, the direction of energy flow is from cold to hot. To achieve this, an external source of energy is needed to drive a **heat pump**.

The principle of a refrigerator is shown in the diagram.

The pipes at the back of a refrigerator are painted black. How does this help to remove energy?

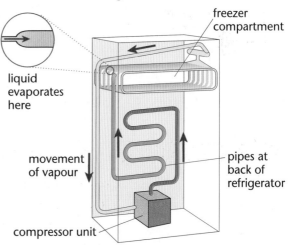

freezer compartment

liquid evaporates here

movement of vapour

pipes at back of refrigerator

compressor unit

A volatile liquid is one that evaporates easily.

- Volatile liquid evaporates as it is forced through a narrow jet inside the refrigerator.
- This evaporation causes cooling and the cold vapour absorbs energy from its surroundings.
- The vapour returns to the outside where it is compressed, causing it to become heated.
- Energy is dissipated to the surroundings as the hot vapour passes through cooling pipes, where it condenses.

This method of heating a building can be used effectively where there is a convenient river to extract the energy from.

Heat pumps are also used to heat buildings by extracting energy from the surroundings. The energy available for heating is more than the energy needed to drive the pump, so this method is both cost-efficient and energy-efficient. Unfortunately, the time of year when most energy is needed is also that when there is least available.

Progress check

1 Hot water enters a central heating radiator at a temperature of 52°C and leaves it at a temperature of 39°C. The specific heat capacity of water is 4200 J kg⁻¹ K⁻¹.

   a Calculate the energy transfer from each kg of water that passes through the radiator.

   b If the rate of flow of water through the radiator is 3.0 kg min⁻¹, calculate the power output of the radiator.

2 The specific latent heat of vaporization of nitrogen is 2.0×10^5 J kg⁻¹. Calculate the energy required to change 0.65 kg of liquid nitrogen into vapour without changing its temperature.

3 Some air is trapped in a sealed syringe. The piston is pushed in and does 450 J of work on the air. While this is happening 300 J of heat flows out of the air.

   a Calculate the change in internal energy of the air.

   b Explain whether the air ends up hotter or colder than it was to start with.

1 a 5.46×10^4 J   b 2.73 kW
2 1.30×10^5 J
3 a +150 J   b hotter as the internal energy has increased.

3.3 Radioactive decay

After studying this section you should be able to:

- *explain how the results of alpha-scattering experiments give evidence for the atomic model*
- *describe the effects of radioactive decay on the nucleus*
- *calculate the half-life of a radioactive isotope from an activity–time graph*

LEARNING SUMMARY

Evidence for the atomic model

AQA A	M1	NICCEA	A2
AQA B	M2	OCR A	A2
EDEXCEL A	M1	OCR B	A2
EDEXCEL B	A2	WJEC	A2

An alpha particle is a positively-charged particle consisting of two protons and two neutrons. The experiments were carried out in a vacuum so that the alpha particles were not scattered by air particles.

The simple atomic model pictures the atom as a tiny, positively-charged nucleus surrounded by negatively-charged particles (electrons) in orbit. Evidence for this model comes from the scattering of alpha particles as they pass through a thin material such as gold foil. The results of these experiments, first carried out under the guidance of Rutherford in 1911, can be summarised as:

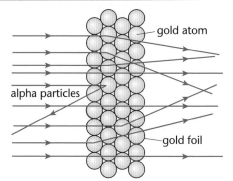

- most of the alpha particles travel straight through the foil with little or no deflection
- a small number are deflected by a large amount
- a tiny number are scattered backwards.

Rutherford concluded that the atoms of gold are mainly empty space, with tiny regions of concentrated charge. This charge must be the same sign as that of alpha particles (positive) to explain the back-scattering as being due to the repulsion between similar-charged objects.

The nucleus

AQA A	M1	NICCEA	A2
AQA B	M2	OCR A	A2
EDEXCEL A	M1	OCR B	A2
EDEXCEL B	A2	WJEC	M2

You will not find El in the periodic table – it is fictitious.

Later experiments showed that there are two types of particle in the nucleus, **protons** and **neutrons**.

> The nucleus of an element is represented as ^A_ZEl.
> Z is the atomic number, the number of protons.
> A is the mass number, the number of nucleons (protons and neutrons).
>
> **KEY POINT**

The element is fixed by Z, the number of protons. A neutral atom has equal numbers of protons in the nucleus and electrons in orbit. Different atoms of the same element can have different values of A due to having more or fewer neutrons. As this does not affect the number of electrons in a neutral atom, the chemical properties of these atoms are the same. They are called **isotopes** of the element. The most common form of carbon, for example, is carbon-12, $^{12}_6$C, which has six protons and six neutrons in the nucleus. Carbon-14, $^{14}_6$C, has the same number of protons but two extra neutrons. These two forms of carbon are isotopes of the same element.

The phrase 'relative to the value' means compared to the actual amount of charge, ignoring the sign.

The charges and relative masses of atomic particles are shown in the table. The masses are in atomic mass units (u), where $1u = 1/12$ the mass of a carbon-12 atom $= 1.661 \times 10^{-27}$ kg. The charges are relative to the value of the electronic charge, $e = 1.602 \times 10^{-19}$ C.

atomic particle	mass	charge
proton	1.01	+1
neutron	1.01	0
electron	5.49×10^{-4}	−1

Radiation all around us

AQA A	A2	NICCEA	A2
AQA B	M2	OCR A	A2
EDEXCEL A	M1	WJEC	A2
EDEXCEL B	M2		

The term 'background radiation' is also used to describe the microwave radiation left over from the 'Big Bang'. This is a different type of radiation to nuclear radiation.

Radioactive decay occurs when an atomic nucleus changes to a more stable form. This is a random event that cannot be predicted. The emissions from these nuclei are collectively called **radioactivity** or **radiation**.

We are subjected to a constant stream of radiation called **background radiation**. Most of this is **natural** in the sense that it is not caused by the activities of people living on Earth. Sources of background radiation include:

- the air that we breathe – radioactive radon gas from rocks can concentrate in buildings
- the ground and buildings – all rocks contain radioactive isotopes
- the food that we eat – the food chain starts with photosynthesis. Radioactivity enters the food chain in the form of carbon-14, an unstable isotope of carbon that is continually being formed in the atmosphere
- radiation from space, called cosmic radiation
- medical and industrial uses of radioactive materials.

The emissions

AQA A	A2	NICCEA	A2
AQA B	M2	OCR A	A2
EDEXCEL A	M1	OCR B	A2
EDEXCEL B	M2	WJEC	A2

The range of a beta-plus particle is effectively zero, since it is annihilated when it collides with an electron.

Alpha radiation is the most intensely ionising and can cause a lot of damage to body tissue. Although it cannot penetrate the skin, alpha emitters can enter the lungs during breathing.

Radioactive emissions are detected by their ability to cause **ionisation**, creating charged particles from neutral atoms and molecules by removing outer electrons. This results in a transfer of energy from the emitted particle, which is effectively absorbed when all its energy has been lost in this way. The four main emissions are **alpha** (α), **beta-plus** (β^+), **beta-minus** (β^-) and **gamma** (γ). Of these, alpha radiation is the most intensely ionising and has the shortest range, with the exception of beta-plus, while gamma radiation is the least intensely ionising and has the longest range.

- In alpha emission the nucleus emits a particle consisting of two protons and two neutrons. This has the same make-up as a helium nucleus.
- Beta-minus emission occurs when a neutron decays into a proton, emitting an electron in the process.
- In beta-plus emission a proton changes to a neutron by emitting a positron.
- A gamma emission is short-wavelength electromagnetic radiation.

Some properties of these radiations are shown in the table.

radioactive emission	nature	charge/e	symbol	penetration	causes ionisation	affected by electric and magnetic fields
alpha	two neutrons and two protons	+2	^4_2He or $^4_2\alpha$	absorbed by paper or a few cm of air	intensely	yes
beta-minus	high-energy electron	−1	$^0_{-1}e$ or $^0_{-1}\beta$	absorbed by 3 mm of aluminium	weakly	yes
beta-plus	positron (antielectron)	+1	$^0_{+1}e$ or $^0_{+1}\beta$	annihilated by an electron		yes
gamma	short-wavelength electromagnetic radiation	none	$^0_0\gamma$	reduced by several cm of lead	very weakly	no

Notice that the electron emitted in beta-minus decay and the positron emitted in beta-plus decay have been allocated the atomic numbers −1 and +1. This is because of the effect on the nucleus when these particles are emitted.

Balanced equations

AQA A	A2	NICCEA	A2
AQA B	M2	OCR A	A2
EDEXCEL A	M1	OCR B	A2
EDEXCEL B	A2	WJEC	A2

Technetium-99 is an artificial isotope that emits gamma radiation only. It is used as a tracer in medicine. The gamma radiation can be detected outside the body and there is less risk of cell damage than with an isotope that also emits alpha or beta radiation.

When a nucleus decays by alpha or beta emission, the numbers of protons and neutrons are changed. Gamma emission does not change the make-up of the nucleus, but corresponds to the nucleus losing excess energy. Gamma emission often occurs alongside alpha and beta emissions, though some artificial radioactive isotopes emit gamma radiation only.

The changes that take place due to alpha and beta emissions are:

- **alpha** – the number of protons decreases by 2 and the number of neutrons also decreases by two
- **beta-minus** – the number of neutrons decreases by one and the number of protons increases by one
- **beta-plus** – the number of neutrons increases by one and the number of protons decreases by one.

In writing equations that describe nuclear decay, both charge (represented by Z) and the number of nucleons (represented by A) are conserved. The table summarises these changes and gives examples of each type of decay.

Check that the equations given as examples are balanced in terms of charge and number of nucleons.

particle emitted	effect on A	effect on Z	example
alpha	−4	−2	$^{226}_{88}Ra \rightarrow ^{222}_{86}Rn + ^{4}_{2}He$
beta-minus	unchanged	+1	$^{14}_{6}C \rightarrow ^{14}_{7}N + ^{0}_{-1}e$
beta-plus	unchanged	−1	$^{11}_{6}C \rightarrow ^{11}_{5}B + ^{0}_{+1}e$

Rate of decay and half-life

AQA A	A2	NICCEA	A2
AQA B	M2	OCR A	A2
EDEXCEL A	M1	OCR B	A2
EDEXCEL B	A2	WJEC	A2

> **KEY POINT**
> The activity, or rate of decay, of a sample of radioactive material is measured in **becquerel** (Bq). An activity of 1 Bq represents a rate of decay of 1 s^{-1}.

Radioactive decay is a random process and the decay of an individual nucleus cannot be predicted. However, given a sample containing large numbers of undecayed nuclei, then statistically the rate of decay should be proportional to the number of undecayed nuclei present. Double the size of the sample and, on average, the rate of decay should also double.

There are only two factors that determine the rate of decay of a sample of radioactive material. They are:

- the radioactive isotope involved
- the number of undecayed nuclei.

Unlike chemical reactions, radioactive decay is not affected by changes in temperature.

> **KEY POINT**
> The relationship between the rate of decay, or activity, of a radioactive isotope is:
> $$activity = \lambda N$$
> where the activity is measured in becquerel, N is the number of undecayed nuclei present and λ is the **decay constant** of the substance. λ has units of s^{-1}.

When the activity of a radioactive isotope is plotted against time, the result is a curve that shows the activity decreasing as the number of undecayed nuclei decreases. A decay curve is shown in the diagram. The shape of the curve is the same for all radioactive substances, but the activities and time scales depend on the size of the sample and its decay constant.

The graph shown is a plot of activity against time. A plot of number of undecayed nuclei against time would be identical but with a different scale on the *y*-axis.

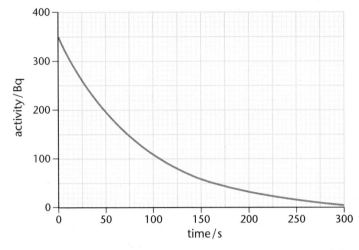

The curve shows **exponential decay**. In an exponential decay curve:

- the rate of change of *y*, represented by the gradient of the curve, is proportional to *y*
- in equal intervals of *x* the value of *y* always changes in the same ratio.

This second point means that it always takes the same time interval for the activity to decrease to a given fraction of any particular value. The time for the activity to halve is a measure of the **half-life** of the substance.

Note that although half-life is defined in terms of the number of undecayed nuclei, it is usually measured as the average time for the activity of a sample to halve.

> **KEY POINT**
>
> The half-life of a radioactive isotope, $t_{\frac{1}{2}}$ is the average time taken for the number of undecayed nuclei of the isotope to halve.

Because radioactive decay is a random process, the results of any experiment to measure activity do not fit the curve exactly and there is some variation in the change of activity when identical samples of the same material are compared. This is why the term *average* is used in the definition of half-life. The values of half-lives range from tiny fractions of a second to many millions of years. Starting with N nuclei of a particular isotope, the number remaining after one half-life has elapsed is $N/2$, after two half-lives $N/4$ and after n half-lives it is $N/2^n$.

There is a relationship between the half-life and the decay constant of any particular radioactive isotope. The shorter the half-life, the greater the rate of decay and the decay constant.

When using this relationship, λ and $t_{\frac{1}{2}}$ must be measured in consistent units, e.g. λ in s^{-1} and $t_{\frac{1}{2}}$ in s.

> **KEY POINT**
>
> The half-life of a radioactive isotope and its decay constant are related by the equation:
> $$\lambda t_{\frac{1}{2}} = \ln 2 = 0.69$$

Progress check

1 **a** What feature of the results of alpha-particle scattering experiments shows that the nucleus is positively charged?

 b Explain how the results of these experiments suggest that the nucleus occupies a very small proportion of the volume of the atom.

2 Strontium-90, $^{90}_{38}Sr$, decays to yttrium, symbol Y, by beta-minus emission. Write a balanced nuclear equation for this decay.

3 Use the graph on this page to determine the values of the half-life and the decay constant for a radioactive isotope that has this decay curve.

3 half-life = 60 s
decay constant = 1.15×10^{-2} s^{-1}

2 $^{90}_{38}Sr \rightarrow ^{90}_{39}Y + ^{0}_{-1}e$

1 **b** Most alpha particles pass through thin gold foil without being deviated.

 a The fact that some alpha particles are back-scattered shows that they are repelled from the nucleus, so they must have the same charge.

3.4 Forces and particles

After studying this section you should be able to:

- distinguish between the groups of fundamental and other particles
- interpret Feynman diagrams
- explain forces in terms of exchange particles
- describe the processes of pair production and pair annihilation

LEARNING SUMMARY

Holding the nucleus together

AQA A	M1	NICCEA	A2
EDEXCEL A	M3C	OCR B	A2
EDEXCEL B	A2		

On the nuclear scale, gravitational forces are so weak that they can be ignored.

The forces that we experience every day can be classified as one of two types:

- **gravitational forces** affect all objects that have mass – they are very weak and have an infinite range
- **electromagnetic forces** also have an infinite range and are much stronger than gravitational forces – they include all forces due to static or moving charges.

The nucleus is very concentrated in terms of both mass and charge. The electrostatic force between the protons is immense and the gravitational attraction is tiny, so if these were the only forces acting, the nucleus would be unstable. In addition to the forces above, there are two other fundamental forces.

The strong force decreases very rapidly with increasing separation of the particles.

- The **strong nuclear force** affects protons and neutrons but not electrons. Its limited range, 1×10^{-15} m, means that it only acts between nucleons that are very close, i.e. next to each other in the nucleus.
- The **weak nuclear force** affects all particles. It has an even shorter range than the strong force, 1×10^{-18} m, and is responsible for beta decay.

It is the strong nuclear force that keeps the nucleus together. The strong attraction between the nucleons balances the electrostatic repulsion.

Fundamental particles

AQA A	M1	NICCEA	A2
AQA B	M2	OCR B	A2
EDEXCEL A	M1	WJEC	A2
EDEXCEL B	A2		

In an inelastic collision, kinetic energy is not conserved.

Electrons are fundamental particles, they cannot be split up. Protons and neutrons are themselves made up of other particles. Evidence for the structure of protons and neutrons comes from **deep inelastic scattering** experiments. In these experiments very-high-energy electrons are fired at nucleons. The electrons are not affected by the strong force and they penetrate the nucleons in an inelastic collision, resulting in the electrons being scattered through a range of angles, with some being back-scattered in the same way that alpha particles can be back-scattered by gold foil.

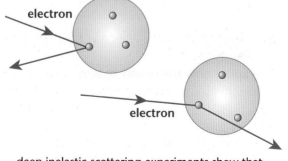

deep inelastic scattering experiments show that
nucleons contain small regions of intense charge

This gives evidence that a nucleon contains small, dense regions of charge that are themselves fundamental particles. These particles are called **quarks**. The diagram above illustrates some of the findings of deep inelastic scattering. In the diagram below each nucleon is shown as containing three quarks.

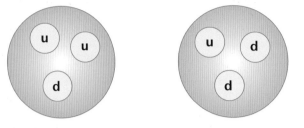

The quarks contained in nucleons

The symbol *e* is used throughout this section to represent the amount of charge on an electron, ignoring the sign. Positive (+) and negative (–) show the sign of a charge.

Of the six types of quark, two occur in protons and neutrons. These are the **up** and **down** quarks. The up quark, symbol u, has a charge of $+2e/3$ and that on the down quark, symbol d, is $-1e/3$. The diagram above shows the quarks in a proton and a neutron.

Particles and their antis

AQA A	M1	EDEXCEL B	A2
AQA B	M2	NICCEA	A2
EDEXCEL A	M3C		

Where a particle has zero charge, for example the neutron, its antiparticle also has zero charge.

For every type of particle there is an antiparticle. An antiparticle:

• has the same mass as the particle, but has the opposite charge if it is the antiparticle of a charged particle
• annihilates its particle when they collide, the collision resulting in energy in the form of gamma radiation or the production of other particles. This process is called **pair annihilation**
• can be created, along with its particle, when a gamma ray passes close to a nucleus. This process is called **pair production**.

Anti-electrons, or **positrons**, are emitted in some types of radioactive decay and can be created when cosmic rays interact with the nuclei of atoms. Like all antiparticles, they cannot exist for very long on Earth before being annihilated by the corresponding particles.

The symbols β^- and β^+ are normally used to refer to electrons and positrons given off as a result of radioactive decay.

> **using symbols for particles**
> An electron is represented by the symbol e^- or β^-.
> The positron can be represented by either of the symbols e^+ and β^+.
> The proton and antiproton are written as p^+ and p^-.
> More generally, a line drawn above the symbol for a particle refers to an antiparticle, so u and $\bar{u}$ (read as u-bar) refer to the up quark and its antiquark.
>
> **KEY POINT**

The greek letter ν is pronounced as 'new'.

The **antineutrino**, $\bar{\nu}$, is a particle of antimatter emitted along with an electron when an nucleus undergoes β^- decay. It is a particle that has almost no mass and no charge. Its corresponding particle the **neutrino**, ν, is emitted with a positron in β^+ decay.

The quark family

AQA A	M1	EDEXCEL B	A2
AQA B	M2	NICCEA	A2
EDEXCEL A	M3C	OCR B	A2

There are six quarks which, together with their antiquarks, make up the family. The name of each quark describes its type, or **flavour**. Some of the properties of quarks are:

The kaons (see later in this section) were described as 'strange' particles when their lifetime turned out to be much longer than expected.

• charge – this is $+2e/3$ or $-1e/3$ for quarks and is always conserved
• baryon number – like charge, this is conserved in all interactions
• strangeness – this property describes the strange behaviour of some particles that contain quarks with strangeness value that is non-zero. It is conserved in strong and electromagnetic interactions but can be changed in weak interactions.

The values of these properties are shown in the table.

The six quarks can be thought of as three pairs, u–d, c–s and t–b.

quark	symbol	charge/e	baryon number, B	strangeness, S
up	u	+2/3	1/3	0
down	d	−1/3	1/3	0
charm	c	+2/3	1/3	0
strange	s	−1/3	1/3	−1
top	t	+2/3	1/3	0
bottom	b	−1/3	1/3	0

The properties of antiquarks are similar to those of the corresponding quark but with the opposite sign, so the antiquark $\bar{s}$ has charge +1/3, baryon number −1/3 and strangeness +1.

The hadron families

AQA A ▸ M1   EDEXCEL B ▸ A2
AQA B ▸ M2   NICCEA ▸ A2
EDEXCEL A ▸ M3C

The **hadrons** and **leptons** are two groups of particles. Hadrons are affected by both the strong and the weak nuclear force, but leptons are affected by the weak force only.

The intermediate vector bosons, responsible for the weak interaction, form another group of particles.

Hadrons are not fundamental particles; they are made up of quarks:
- **baryons** are made up of three quarks
- **mesons** consist of a quark and an antiquark.

The familiar baryons are the proton, whose quark structure is uud, and the neutron, udd. There are other baryons which have different combinations of quarks but all are unstable and have short lifetimes. All baryons have a baryon number +1, as each quark contributes a baryon number of +1/3.

Two families of mesons are the **pi-mesons**, or **pions**, and the **kaons**. Each family consists of three particles with charges +1, −1 and 0 along with their antiparticles. Because a meson consists of a quark and an antiquark, it has baryon number 0.

The table shows the structure and some properties of the kaons and pions.

particle	structure	charge/e	baryon number, B	strangeness, S
π^0	$u\bar{u}$ or $d\bar{d}$	0	0	0
π^+	$u\bar{d}$	+1	0	0
π^-	$\bar{u}d$	−1	0	0
K^0	$d\bar{s}$	0	0	+1
K^+	$u\bar{s}$	+1	0	+1
K^-	$\bar{u}s$	−1	0	−1

The leptons

AQA A ▸ M1   EDEXCEL B ▸ A2
AQA B ▸ M2   NICCEA ▸ A2
EDEXCEL A ▸ M3C

Remember, a fundamental particle cannot be split up into other, simpler particles.

Leptons are fundamental particles. There are three families of leptons:
- the electron and its antiparticle the positron, e^- and e^+, together with the neutrinos given off in β^- and β^+ decay. These neutrinos are called the electron-antineutrino, $\bar{v}_e$, and the electron-neutrino, v_e, to distinguish them from the neutrinos associated with the other leptons
- the muons, μ^- and μ^+, together with their neutrinos v_μ and $\bar{v}_\mu$. Muons have 200 times the mass of electrons, but are unstable, decaying to an electron and a neutrino
- the tauons, τ^- and τ^+ and their neutrinos v_τ and $\bar{v}_\tau$. The tauons are the most massive leptons, having twice the mass of a proton. Like the muons, they are unstable.

Cosmic rays create pions in the upper atmosphere due to collisions with nuclei. These pions decay, charged pions producing muons and their neutrinos and uncharged pions decaying to gamma radiation.

Muons are produced by cosmic rays. Their properties are similar to those of electrons and positrons but they are much more massive. Tauons are the elephants of the lepton tribe; they are produced by the interaction of high-energy electrons and positrons.

Exchanging particles

The concept of a **field** is used to describe the effects of gravitational and electromagnetic forces and how they vary with distance. This is a useful concept that enables the size and direction of forces to be calculated, but it does not explain what causes them.

The interaction between particles that results in attractive and repulsive forces is due to a continual exchange of other particles. These **exchange particles** have a very short lifetime and owe their existence to borrowed energy, so they are often referred to as **virtual particles**.

There are different exchange particles associated with each of the four fundamental forces. For electromagnetic forces the exchange particles are photons, short bursts of electromagnetic radiation.

> A photon is a 'packet' of electromagnetic radiation. If you are unfamiliar with the concept of a photon, refer to chapter 4.

When two electrons approach each other they exchange virtual photons. The closer the electrons are to each other, the shorter the wavelength of the virtual photons exchanged.

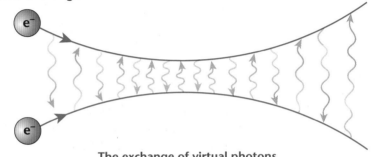

The exchange of virtual photons between two interacting electrons

This exchange of virtual photons results in the electrons moving away from each other in the same way as two ice skaters on a collision course would move away from each other if they were to keep throwing things at each other!

The diagram above represents the exchange of virtual photons between two interacting electrons.

> Richard Feynman was an American physicist. He is famous for his books *Lectures on Physics* and for his way of explaining physics concepts by using diagrams.

Interactions between particles can be represented by **Feynman diagrams**. In these diagrams:

- straight lines are used for the interacting particles
- wavy lines represent the exchange particles
- the straight lines extend beyond the diagram, showing the existence of these particles before and after the interaction
- the wavy lines are contained within the diagram, showing that the exchange particles are short-lived.

The diagram opposite is a Feynman diagram for the interaction between two electrons shown in the diagram above. The arrows do not represent directions of travel; arrows pointing in show the particles before the interaction and those pointing out show the particles after the interaction. The symbol γ is used to represent the exchange of virtual photons.

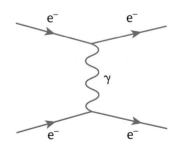

> The same symbol, g, is used for both gluons and gravitons.

The exchange particles for gravitational forces are called **gravitons**, symbol g, though these have not yet been observed.

The strong force

It is the strong force that holds nucleons together in a nucleus and holds quarks together in a nucleon. The exchange particles between quarks are called **gluons**, symbol g. The quarks that make up a proton or neutron are constantly exchanging gluons as they interact with each other.

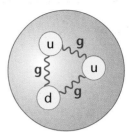

The exchange of gluons between the quarks in a proton

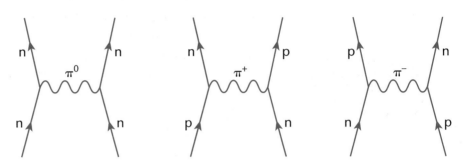

Interactions between nucleons

> Since there is no change in charge in the interactions between two neutrons or two protons, only neutral particles can be exchanged.

Pions, or pi-mesons, are responsible for the strong force between nucleons. Two protons or two neutrons exchange pi-zero, π^0, particles, but any one of the three pions can be exchanged during an interaction between a proton and a neutron. Feynman diagrams for some of the possible interactions are shown above.

Notice that charge and baryon number are conserved at each junction of the diagrams.

The weak force

> The intermediate vector bosons form another group of particles, separate from the hadrons and leptons.

Weak interactions involve the exchange of one of three particles called **intermediate vector bosons**. Like the pi-mesons, their symbols, W^+, W^- and Z^0 indicate the charge on each boson. The W^+ boson carries charge $+e$ and the W^- boson carries charge $-e$.

- The W^+ boson transfers charge $+e$; it is exchanged in an interaction between a neutrino and a neutron, resulting in an electron and a proton.
- The W^- boson is responsible for β^- decay of a neutron.
- Weak interactions where there is no transfer of charge involve the Z^0 boson, for example a **collision** between an electron and an antineutrino.

Feynman diagrams for these interactions are shown below.

> Can you interpret these Feynman diagrams? Try to describe the interaction that each one represents.

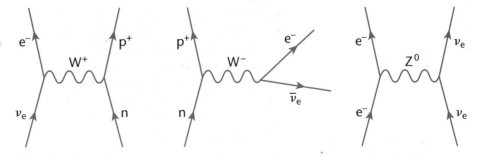

In two of these examples the type or flavour of a quark in the neutron has been changed from d to u by the weak interaction involving a charge-carrying boson. In β⁻ decay a down quark emits a W⁻ particle which decays to an electron and its antineutrino. This is shown in the diagram below.

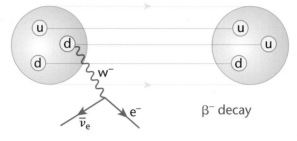

β⁻ decay

Progress check

1 Describe the differences between:
   a Hadrons and leptons.
   b Baryons and mesons.

2 The diagram represents the decay of a meson.

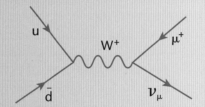

   a Which meson is decaying?
   b What is the exchange particle?
   c What particles are produced as a result of the decay process?

3 What change takes place in a proton in β⁺ decay?

3 A u quark changes to a d quark, changing the quark structure of the nucleon from uud to udd.
   c The μ⁺ muon and its neutrino.
   b The intermediate vector boson W⁺.
   2 a π⁺.
   b Baryons contain 3 quarks. Mesons contain a quark and an antiquark.
   Leptons are fundamental particles.
   1 a Hadrons are made up of quarks.

3.5 Stability and the nucleus

After studying this section you should be able to:

- use the N–Z curve to explain how the type of decay depends on the position of an unstable nucleus relative to the region of stability
- explain why energy is released in nuclear decay
- describe the similarities and differences in the energy spectra for α, β and γ emission

LEARNING SUMMARY

How big?

AQA A   A2   NICCEA   A2
EDEXCEL A   M3C

Atoms and nuclei are not hard, rigid particles with a fixed shape and size. Evidence from alpha-scattering experiments shows that the diameter of a nucleus is approximately one ten-thousandth (1×10^{-4}) that of the atom and the nuclear volume is 1×10^{-12} that of the atomic volume. As almost all the atomic mass is in the nucleus, it must be extremely dense.

Electron diffraction experiments allow more precise estimates of the size of nuclei to be made. The results of these experiments show that:

Unlike their nuclei, the atoms of different elements can have very different densities.

- all nuclei have approximately the same density, about 1×10^{17} kg m^{-3}
- the greater the number of nucleons, the larger the radius of the nucleus
- the nucleon number, A, is proportional to the cube of the nuclear radius, r.

An alternative way of writing this last point is:

- the nuclear radius, r, is proportional to the cube root of the nucleon number, A.

> **KEY POINT**
>
> The relationship between nucleon number, A, and nuclear radius, r, is:
> $$r = r_0 A^{1/3}$$
> where r_0 has the value 1.2×10^{-15} m.

This relationship means that if the nuclei of ^{64}Cu and ^{32}S are compared, the radius of the copper nucleus is $2^{1/3} = 1.26$ times that of the sulphur, as the nucleon number of copper is twice that of sulphur. Similarly, the copper nucleus has a radius which is $4^{1/3} = 1.59$ times that of ^{16}O.

Stable and unstable nuclei

AQA A   A2   OCR B   A2
EDEXCEL A   M1, M3C

Carbon-11, carbon-12 and carbon-14 are three isotopes of carbon. Of these, only carbon-12 is stable. It has equal numbers of protons and neutrons. The graph shows the relationship between the number of neutrons (N) and the number of protons (Z) for stable nuclei.

Remember, isotopes of an element all have the same number of protons but different numbers of neutrons in the nucleus.

It can be seen from this graph that the condition for a nucleus to be stable depends on the number of protons:

- for values of Z up to 20, a stable nucleus has equal numbers of protons and neutrons
- for values of Z greater than 20, a stable nucleus has more neutrons than protons.

For stable nuclei with more than 20 protons, the neutron:proton ratio increases steadily to a value of around 1.5 for the most massive nuclei.

Unstable nuclei above the stability line in the graph are **neutron-rich**; they can

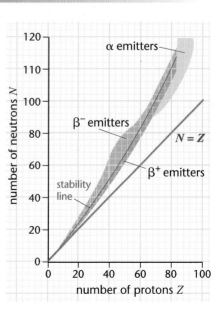

become more stable by decreasing the number of neutrons. They decay by β^- emission; this leads to one less neutron and one extra proton and brings the neutron–proton ratio closer to, or equal to, one. An example is:

$$^{24}_{11}\text{Na} \rightarrow {}^{24}_{12}\text{Mg} + {}^{0}_{-1}\text{e}$$

Unstable nuclei below the stability line decay by β^+ emission; this increases the neutron number by one at the expense of the proton number. An example is:

$$^{11}_{6}\text{C} \rightarrow {}^{11}_{5}\text{B} + {}^{0}_{+1}\text{e}$$

The emission of an alpha particle has little effect on the neutron–proton ratio for isotopes that are close to the $N=Z$ line and is confined to the more massive nuclei. For these nuclei, emission of an alpha particle changes the balance of the proton–neutron ratio in the favour of the neutrons. In the decay of thorium-228 shown below, the neutron–proton ratio increases from 1.53 to 1.55.

$$^{228}_{90}\text{Th} \rightarrow {}^{224}_{88}\text{Ra} + {}^{4}_{2}\text{He}$$

> The $N=Z$ line corresponds to a neutron–proton ratio of 1. As an alpha particle consists of two neutrons and two protons, its emission would hardly affect a neutron–proton ratio that is nearly 1.

Decay chains and dating

EDEXCEL A ▶ M3C

> There are four decay chains, starting with actinium, neptunium, thorium and uranium. All except the neptunium chain are found naturally in rocks.

The decay of a nucleus may result in another unstable nucleus which in turn decays and may result in another unstable nucleus. This process is called a **decay chain** or **radioactive series** and it ends in the formation of a nucleus that is stable. The thorium decay chain starts with thorium-232 and ends with lead-208. The beginning of the chain is shown below:

$$^{232}_{90}\text{Th} \rightarrow {}^{228}_{88}\text{Ra} + {}^{4}_{2}\text{He} \rightarrow {}^{228}_{89}\text{Ac} + {}^{0}_{-1}\text{e} \rightarrow {}^{228}_{90}\text{Th} + {}^{0}_{-1}\text{e} \rightarrow {}^{224}_{88}\text{Ra} + {}^{4}_{2}\text{He} \rightarrow$$

A decay chain can be represented on a chart of nucleon number, A, against proton number, Z. The diagram below shows the complete decay chain for thorium-232. On this chart:

- alpha decay is shown by a movement of two units to the left and four units down
- beta-minus decay involves a movement of one unit to the right
- the chain branches at bismuth-212, which has two possible modes of decay.

> Bismuth-212 decays by both alpha and beta-minus emission.

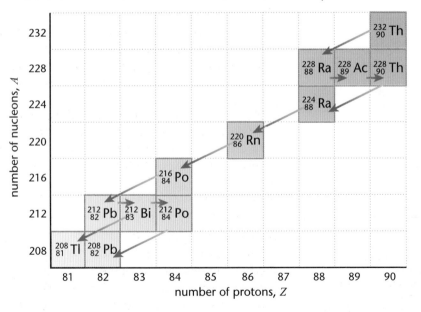

In a similar chain, the decay of uranium-238 ends with the stable isotope lead-206.

> This assumes that there was no lead-206 present when the rock was formed.

In a similar chain, the decay of uranium-238 ends with the stable isotope lead-206. Uranium-238 decays with a half-life of 4.5×10^6 years, the intermediate isotopes having relatively short half-lives. This allows the age of rocks to be estimated by comparing the numbers of uranium-238 atoms and lead-206 atoms present in the rock. A ratio of 1:1 indicates that the age of the rock is the same as the half-life of uranium-238. The oldest rocks found have this ratio of uranium-238 to lead-206 present, showing that the age of the Earth is approximately 4.5×10^9 years.

In **radiocarbon dating** a similar technique is used. Radiocarbon dating compares the concentrations of carbon-14 atoms and carbon-12 atoms in a sample of material. Living things maintain a constant level of carbon-14 which then decays after death. The lower the concentration of carbon-14, the longer the time since death. This method of dating can only be used for objects made from once-living material. The half-life of carbon-14 is 5730 years, so it is not suitable for dating objects that are relatively new (less than five hundred years since death) or very old (more than one hundred thousand years since death).

A conservation problem

AQA A	A2	NICCEA	A2
AQA B	A2	OCR A	A2
EDEXCEL A	M3C	OCR B	A2
EDEXCEL B	A2	WJEC	A2

A carbon-12 nucleus consists of six protons and six neutrons. The mass of the atom is precisely 12u, by definition, so after taking into account the mass of the electrons, that of the nucleus is 11.9967 u. The mass of the constituent neutrons and protons is:

$$6\,m_p + 6\,m_n = 6(1.0073\ u + 1.0087\ u) = 12.0960\ u$$

The nucleus has less mass than the particles that make it up. This appears to contravene the principle of conservation of mass. Einstein established that *energy has mass*. The mass that you gain due to your increased energy when you walk upstairs is infinitesimally small, but at a nuclear level this mass cannot be ignored. Any change in energy is accompanied by a change in mass, and vice versa.

Einstein's equation $E = mc^2$ gives the method for working out how much mass is associated with energy. Try using it to work out the increase in your mass when you walk upstairs.

When dealing with the nucleus and nuclear particles, energy and mass are so closely linked that their equivalence, and that of their units, has been established:

> **KEY POINT**
>
> 1 u = 930 MeV
> where 1 eV (one electron volt) is the energy transfer when an electron moves through a potential difference of 1 volt.

An electronvolt is a tiny amount of energy compared to the joule, which is too large a unit to use on an atomic scale. 1 eV = 1.60×10^{-19} J.

Using this relationship, the separate conservation rules regarding mass and energy can be combined into one so that (mass + energy) is always conserved in nuclear interactions.

To split a nucleus up into its constituent nucleons requires energy. It follows that a nucleus has less energy than the sum of the energies of the corresponding number of free neutrons and protons. So the fact that a nucleus has less energy than its nucleons would have in isolation, means that it also has less mass.

> **KEY POINT**
>
> The difference between the sum of the masses of the individual nucleons and the mass of the nucleus is called the **mass defect** or **nuclear binding energy**. It represents the energy required to separate a nucleus into its individual nucleons.

In the case of carbon-12 the mass defect, or nuclear binding energy, is equal to 0.093 u = 89.3 MeV.

The greater the binding energy per nucleon, the more energy (per nucleon) is required to split the nucleus up, giving it more stability.

As would be expected, the greater the number of nucleons, the greater the binding energy. The figure shows how the **binding energy per nucleon** varies with nucleon number. The most stable nuclei have the greatest binding energy per nucleon.

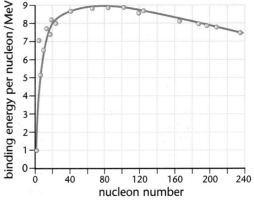

111

Energy spectra

AQA A ▶ A2   EDEXCEL A ▶ M3C

A nucleus can only decay naturally to a state where it has less energy.

The kinetic energy of the decay products includes that of the recoiling nucleus. This is most significant in the case of alpha decay.

When a nucleus decays, the daughter nucleus has less mass and so more binding energy than the parent. The energy difference is the **decay energy**, the energy released by the nucleus as a result of its decay. This energy is transferred to:

- kinetic energy of the decay products
- energy of a gamma ray, if one is emitted
- energy of a neutrino or antineutrino, in the case of β emission.

There are different patterns in the energies of alpha, beta and gamma emissions.

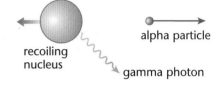

Energy being removed from a nucleus in alpha emission

recoiling nucleus

alpha particle

gamma photon

- The energy of alpha particles depends on the source; some sources emit alpha particles that all have the same energy, other sources emit alpha particles with two or more possible energies.
- The energy of beta particles varies continuously over a range from zero up to a maximum value that depends on the source.
- In gamma emission, each source emits a line spectrum; the photons have only one or a small number of energies.

The energies of the emitted particles are different for different sources. The shorter the half-life, the greater the energy of the emissions.

These cloud chamber tracks are from a source that emits alpha particles of two separate energies.

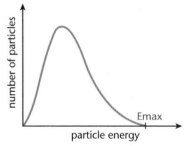

number of particles

particle energy

Emax

The range of energies in beta emission

Whilst alpha and gamma emissions correspond to certain energy values, in beta emission most of the particles are emitted with less than the maximum energy available. This led to the discovery of the neutrino, ν, and antineutrino, ν̄. The antineutrino is a particle emitted along with the electron in β⁻ decay and a neutrino is emitted along with a positron in β⁺ decay. The energies of these particles account for the difference between the energy released by the nucleus and the kinetic energies of the beta particle and the recoiling nucleus together with the energy of any gamma ray photon that is emitted.

Remember that gamma emission is usually due to a nucleus being left in an excited state following the emission of an alpha or beta particle.

Progress check

1 Tritium, $_1^3$H, is a radioactive isotope of hydrogen. Use the *N–Z* graph to explain how it decays and to write an equation for the decay.

2 The nuclear mass of oxygen-16 is 15.9905 u.
  a Use the data on page 111 to calculate the mass defect in u.
  b Calculate the energy equivalent in MeV.

3 a In what way is the energy spectrum of beta emissions different to that of alpha and gamma emissions?
  b Explain how this is accounted for.

b This is explained by the energy of the neutrino or antineutrino that accompanies a beta particle.

3 a The beta emission spectrum is continuous, whereas those of alpha and gamma only contain certain values of energy.

b 128 MeV

2 a 8(1.0073 u + 1.0087 u) = 16.1280 u − 15.9905 u = 0.1375 u.
   $_1^3$H→$_2^3$He+$_{-1}^0$e

1 As tritium lies above the *N–Z* line, it decays by beta-minus emission.

Sample question and model answer

An ice cube of mass 12.0 g at a temperature of –9°C is placed in a glass containing 100 g of water at 20°C. The mass of the glass is 250 g.
Calculate the equilibrium temperature when the ice has melted. [8]

Specific heat capacities: ice $2.1 \text{ J g}^{-1}\text{ K}^{-1}$
water $4.2 \text{ J g}^{-1}\text{ K}^{-1}$
glass $0.67 \text{ J g}^{-1}\text{ K}^{-1}$

Specific latent heat of ice = 340 J g^{-1}

At equilibrium, the ice has changed into water and all the water and the glass are at the same temperature. The glass and the original water have lost energy, and the melted ice has gained energy.
Assuming no energy transfer to or from the surroundings, these gains and losses must be equal. The equation opposite relates the energy gained by the ice to that lost by the glass and water.

energy gained by ice = energy lost by glass and water   1 mark

There are three separate components in working out the energy gained by the ice. These are:
• energy to warm the ice to 0°C
• energy to melt the ice
• energy to warm the melted ice to the final temperature.

energy gained by ice   = energy to warm the ice to 0°C + energy to melt the ice + energy to heat the melted ice to x°C   1 mark

$= m_{ice} \times c_{ice} \times \Delta\theta_{ice} + m_{ice} \times L_{ice} + m_{ice} \times c_{water} \times \Delta\theta_{water}$ 1 mark

This total energy is balanced by that lost by both the glass and the water.

energy lost by glass and water
$= m_{water} \times c_{water} \times \Delta\theta_{water} + m_{glass} \times c_{glass} \times \Delta\theta_{glass}$ 1 mark

Having established the physical principle, the numerical values are now substituted into the equation. It is necessary to have a symbol for the equilibrium temperature.

If the equilibrium temperature is x°C:
$12.0 \text{ g} \times 2.1 \text{ J g}^{-1}\text{ K}^{-1} \times 9 \text{ K} + 12.0 \text{ g} \times 340 \text{ J g}^{-1} + 12.0 \text{ g} \times 4.2 \text{ J g}^{-1}\text{ K}^{-1} \times x \text{ K} = 100 \text{ g} \times 4.2 \text{ J g}^{-1}\text{ K}^{-1} \times (20 - x) \text{ K} + 250 \text{ g} \times 0.67 \text{ J g}^{-1}\text{ K}^{-1} \times (20 - x) \text{ K}$ 2 marks

This equation now has to be solved to find the value of x.

$226.8 \text{ J} + 4080 \text{ J} + 50.4 \times x \text{ J} = 420 \times (20 - x) \text{ J} + 167.5 \times (20 - x) \text{ J}$
$= 11\ 750 \text{ J} - 587.5 \times x \text{ J}$ 1 mark

Now collect the terms involving numbers on one side of the equation, and those involving x on the other side to give:

$637.9x = 7443$ so $x = 11.7$

The final temperature of the mixture is 11.7°C.   1 mark

Practice examination questions

1 A sample of an ideal gas has a volume of 1.50×10^{-4} m³ at a temperature of 22°C and a pressure of 1.20×10^5 Pa.

(a) Calculate the temperature of the gas in kelvin. [1]

(b) Calculate the number of moles in the sample.
$R = 8.3$ J mol⁻¹. [3]

(c) If the volume of the gas remains constant, calculate the temperature at which the pressure of the gas is 2.00×10^5 Pa. [3]

(d) (i) Calculate the mean kinetic energy of the gas particles at a temperature of 22°C. The Boltzmann constant, $k = 1.38 \times 10^{-23}$ J K⁻¹. [3]

(ii) At what temperature would this mean kinetic energy be doubled? Explain how you arrive at your answer. [2]

2 Calculate the energy released when 1.8 kg of steam at 100°C is changed into water at 85°C. [3]
specific latent heat of vaporisation of water = 2.25×10^6 J kg⁻¹
specific heat capacity of water = 4.20×10^3 J kg⁻¹ K⁻¹

3 The diagram shows how an electric immersion heater can be used to measure the specific heat capacity of a metal block.

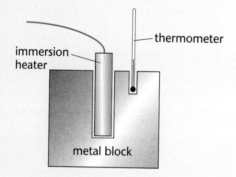

(a) (i) What is the advantage of heating the block through a temperature difference of 10 K rather than 1 K? [1]

(ii) What is the advantage of heating the block through a temperature difference of 10 K rather than 100 K? [1]

(iii) Explain whether you would expect the experiment to give a higher value or a lower value than that given in a data book. [2]

(iv) Suggest one way of improving the reliability of the result. [1]

(b) Use the data to calculate a value for the specific heat capacity of the metal. [3]

heater power = 24 W
time heater switched on = 350 s
mass of block = 0.80 kg
initial temperature = 22°C
final temperature = 36°C.

4   The first law of thermodynamics can be written as:
$$\Delta U = Q + W$$

(a)   State the meaning of each symbol in the equation.   [3]

(b)   As a gas is being compressed, 4500 J of work are done on it and 4800 J of heat are removed from it. Explain what happens to the temperature of the gas as a result of this.   [2]

5   The diagram shows a heat engine.

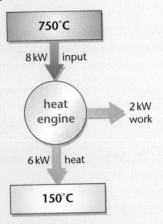

(a)   Calculate the efficiency of the heat engine.   [2]

(b)   What is the maximum efficiency of the engine?   [2]

(c)   Suggest one reason why the efficiency of the heat engine is less than its theoretical maximum.   [1]

6   The diagram shows some possible outcomes when alpha particles are fired at gold foil.

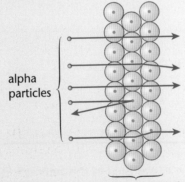

atoms in gold foil

(a)   Explain how these outcomes show that:
   (i)   the nucleus is positively charged   [2]
   (ii) most of the atom is empty space.   [1]

(b)   In **deep inelastic scattering** electrons are used to penetrate the nucleus.
   (i)   What is meant by 'deep inelastic scattering'?   [2]
   (ii)   Explain why electrons are able to probe further into the nucleus than alpha particles can.   [2]
   (iii) Explain why the proton and neutron are not fundamental particles.   [1]
   (iv)   How does deep inelastic scattering provide evidence for quarks?   [1]
   (v)   Describe the quark structure of a proton and a neutron.   [2]

Practice examination questions (continued)

7 1.00 g of carbon is obtained from the wood of a living tree.
(a) Calculate the number of carbon atoms present in 1.00 g, assuming that all the carbon atoms are carbon-12.
The Avagadro constant, $N_A = 6.02 \times 10^{23}$ mol^{-1}.
The mass of 1 mole of carbon-12 is 12 g precisely. [2]

(b) Natural carbon consists of approximately:
99% carbon-12
1% carbon-13
1×10^{-10} % carbon-14.

  (i) Explain whether the actual number of carbon atoms in a 1.00 g sample is likely to be significantly different from the answer to (a). [2]
  (ii) Estimate the number of carbon-14 atoms present in the sample. [1]

(c) The rate of decay of carbon-14 in the sample is 0.25 Bq.
  (i) Calculate the value of λ, the decay constant. [2]
  (ii) Use this value to estimate the half-life of carbon-14. [2]

8 For nuclei with up to 20 protons, a stable nucleus has approximately equal numbers of protons and neutrons.
(a) Explain how, for nuclei with less than 20 protons, the type of decay undergone by unstable nuclei depends on the neutron–proton ratio. [2]

(b) Describe how the decay of $^{13}_{6}$C results in the formation of a more stable nucleus. [2]

(c) How does the neutron–proton ratio change for more massive stable nuclei? [2]

(a) How does the structure of a **meson** differ from that of a **baryon**? [2]

(b) The Feynman diagram illustrates the interaction between two protons.

9

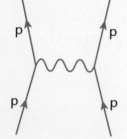

The exchange particle is a pion.
(i) Which pion is exchanged in this interaction? [1]
(ii) Explain why it is this pion rather than a different one. [2]

Chapter 4

Electromagnetic radiation

The following topics are covered in this chapter:

- *What is a wave?*
- *Simple wave properties*
- *More complex wave behaviour*
- *Wave properties across the spectrum*

4.1 What is a wave?

After studying this section you should be able to:

- *describe the difference between a longitudinal and a transverse wave*
- *use the wave equation to calculate speed, frequency and wavelength*
- *explain the meaning of phase and phase difference*

LEARNING SUMMARY

Wave motion

AQA A	A2	NICCEA	M2
AQA B	M2	OCR A	M3
EDEXCEL A	A2	OCR B	M1
EDEXCEL B	M1	WJEC	M1

The waves on a water surface are almost transverse, the particles move in an elliptical path.

Waves are used to transfer a signal or energy. They do this without an accompanying flow of material, although some waves, such as sound, can only be transmitted by the particles of a substance.

All waves consist of **vibrations** or **oscillations**. Sound and other compression waves are classified as **longitudinal** because the vibrations of the particles carrying the wave are along, or parallel to, the direction of wave travel. All electromagnetic waves are **transverse**; the vibrations of the electric and magnetic fields in these waves are at right angles to the direction of wave travel.

Waves transmitted through the body of water, or any other liquid, are longitudinal.

The diagram shows a longitudinal wave being transmitted by a spring and a transverse wave on a rope.

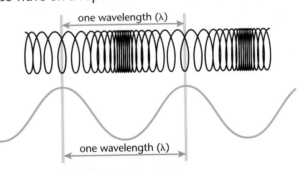

one wavelength (λ)

one wavelength (λ)

Wave measurements

AQA A	A2	NICCEA	M2
AQA B	M1, M2	OCR A	M3
EDEXCEL A	A2	OCR B	M1
EDEXCEL B	M1	WJEC	M1

Most waves that you come across everyday are progressive. Waves that are not progressive are discussed in the section 'more complex wave behaviour'.

These measurements apply to all **progressive** waves. A progressive wave is one that has a profile that moves through space.

- **Wavelength** (symbol λ) is the length of one complete cycle; a compression (squash) and rarefaction (stretch) in the case of a longitudinal wave, a peak and a trough in the case of a transverse wave – see diagram above.
- **Amplitude** (symbol a) is the maximum displacement from the mean position, see diagram on page 118.
- **Frequency** (symbol f) is the number of vibrations per second, measured in hertz (Hz).
- **Speed** (symbol v) is the speed at which the profile moves through space.
- **Period** (symbol T) is the time taken for one vibration to occur. It is related to frequency by the equation $T = 1/f$.

> **KEY POINT**
>
> For all waves, the wavelength, speed and frequency are related by the equation:
>
> $$\text{speed} = \text{frequency} \times \text{wavelength}$$
> $$v = f \times \lambda$$

Wave graphs

AQA A	A2	NICCEA	M2
AQA B	M2	OCR A	M3
EDEXCEL A	A2	OCR B	M1
EDEXCEL B	M1	WJEC	M1

Both transverse and longitudinal waves can be represented by graphs of displacement against position. The position along a wavefront can be measured in two ways:

- as a distance from a point in space
- as an angle.

In angular measure, one complete cycle of the wave is represented by an angle of 360°. This is shown in the diagram.

A displacement–position or displacement–time graph does not distinguish between a transverse wave and a longitudinal one.

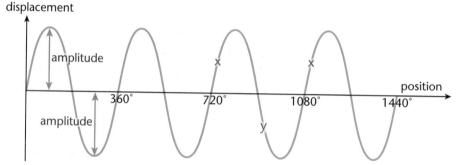

Using angular measure is a convenient way of comparing the **phase** of different parts of a wave. Two points on a wave are **in phase** if they have the same displacement and velocity. For progressive waves, this is only true for points exactly n wavelengths or $360n°$ apart, where n is an integer. Points marked x on the diagram are in phase.

Try substituting values of n, for $n = 0, 1, 2$, etc. into the expression for 'an odd number of half wavelengths'. This will help you to understand the meaning of the phrase.

The point marked **y** is exactly **out of phase** with the points **x** as it has the opposite displacement and its velocity has the same value but in the opposite direction. This applies to any two points separated by an odd number of half wavelengths (i.e. 1/2, 3/2, 5/2 wavelengths, etc.). This can be expressed algebraically as $(2n+1)\lambda/2$, where n is an integer.

The **phase difference** between two points on a wave describes their relative displacement and velocity and is normally expressed in degrees or wavelengths. A phase difference of $\lambda/2$ or 180° describes two points separated by an odd number of half wavelengths (i.e. exactly out of phase), while a phase difference of $\lambda/4$ or 45° compares two points that are separated by one-quarter, five-quarters, nine-quarters of a wavelength, etc.

A common error is to misinterpret a CRO display as a graph of displacement against distance.
This leads to a horizontal measurement, which represents time, being interpreted as a distance.

A **cathode ray oscilloscope** (CRO) can be used to display a different type of graph that represents wave motion. Rather than showing the displacement along a wave, the CRO plots the displacement of one point on the wave against time. The display on the CRO looks the same as that in the diagram, but the horizontal axis represents time rather than position on the wave.

To measure the **period** and **frequency** of a wave from a CRO display, the time–base control, which controls the rate at which the dot sweeps the screen horizontally, needs to be in the 'cal' (calibration) position. The diagram opposite shows a CRO display of a sound wave, with the time–base set to 2 ms cm⁻¹.

One cycle of the wave occupies a distance of 8 cm on the screen, so the period, $T = 8 \times 2$ ms = 16 ms = 1.6×10^{-2} s.

The frequency of the wave, $f = 1/T = 1 \div 1.6 \times 10^{-2}$ s = 62.5 Hz.

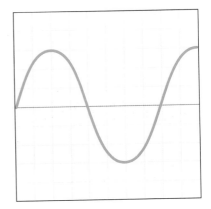

Progress check

1 Describe the difference between a longitudinal and a transverse wave and give one example of each.

2 Calculate the frequency of a VHF radio broadcast that has a wavelength of 2.75 m. The speed of radio waves, $c = 3.00 \times 10^8$ m s^{-1}.

3 Two points on a wavefront have a phase difference of 180°. Describe their relative displacement and velocity.

3 The points have the same amount of displacement but in opposite directions, i.e. the displacement of each one is the negative of the displacement of the other one. Similarly, the velocities are equal in size but opposite in direction.

2 1.09×10^8 Hz.

1 In a longitudinal wave the vibrations are parallel to the direction of travel, e.g. sound or any other compression wave.
In a transverse wave the vibrations are at right angles to the direction of travel, e.g. light or any other electromagnetic wave.

4.2 Simple wave properties

After studying this section you should be able to:

- *explain the meaning of the terms* polarisation *and* intensity
- *use the relationships between refractive index, speeds of light in media and angles of incidence and refraction*
- *describe how optical fibres are used in communications*

Polarisation

AQA A	A2	NICCEA	M2
AQA B	M2	OCR A	M3
EDEXCEL A	A2	OCR B	M1
EDEXCEL B	M2	WJEC	M1

Polarising material is used in some types of sunglasses and in filters for camera lenses. It cuts out reflected light from a flat surface such as an expanse of water.
Can you work out how it does this?

In order to receive a television or radio broadcast, the receiving aerial has to be lined up with that of the transmitter. This is because radio waves broadcast from aerials are **polarised**, the vibrations are only in one plane. For radio and television broadcasts the **plane of polarisation** is usually either vertical or horizontal.

Light waves are not normally polarised, and nor are radio waves from stars. In an unpolarised wave the vibrations are in all planes at right angles to the direction of travel. Polarisation does not apply to longitudinal waves.

The diagram shows the vibrations in polarised and unpolarised waves. Light waves can become polarised as they pass through some materials and they are partially polarised when reflected.

Intensity

AQA B	M2	OCR A	M3
EDEXCEL A	M1	OCR B	M2
EDEXCEL B	M1	WJEC	M1

The signal strength of a radio broadcast decreases the further away the receiver is from the transmitting aerial. In a similar way, a sound appears fainter and a light seems dimmer further away from their origins. This is due to energy becoming spread over a wider area as the wave travels away from its source.

The **intensity** with which a wave is received depends on the power incident on the area of the detector.

> **KEY POINT**
>
> Intensity = power per unit area measured at right angles to the direction of travel.
>
> $$I = P/A$$
>
> Intensity is measured in W m^{-2}.

A point source such as a star radiates energy in all directions. In this case the variation of intensity with distance follows an inverse square pattern, with:

$$I \propto 1/r^2$$

The diagram opposite shows the sound spreading out from a loudspeaker. The ear that is further away detects a quieter sound because the power is spread over a wider area, so less enters the ear.

Reflecting light

AQA B	M2	OCR B	M1
OCR A	M3		

The image in a mirror is virtual, upright, the same size as the object and the same distance behind the mirror as the object is in front.

Mirrors form images because they reflect light in a regular way. When light is reflected at a mirror:

- the angles of incidence and reflection are equal
- the incident light, reflected light and the normal all lie in the same plane.

These are illustrated in the diagram opposite, which serves as a reminder that the angles of incidence and reflection are measured relative to the normal line, a line drawn at right angles to the surface.

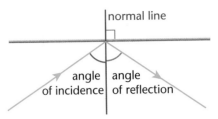

The second rule merely states that the event can be captured on a flat sheet of paper.

Changing speed and direction

AQA A	M1	OCR A	M3
AQA B	M2	OCR B	M1
EDEXCEL B	M1, M2	WJEC	M1
NICCEA	M2		

Refraction also causes virtual images. Images due to refraction are similar to the object, the only difference being that they are closer to the observer.

When waves cross a boundary between two materials there is a change of speed, called **refraction**. If the direction of wave travel is at any angle other than along the normal line, this change in speed causes a change in direction. The diagram opposite shows what happens to light passing through glass. Note that there is always some reflection at a boundary where refraction takes place.

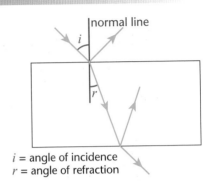

i = angle of incidence
r = angle of refraction

The change in speed is described by the **refractive index**.

> **KEY POINT**
>
> The refractive index of a material, n, is the ratio of the speed of light in a vacuum to the speed of light in the material.
> $$n = c_v \div c_m$$
> where c_v is the speed of light in a vacuum and c_m is the speed of light in the material.
> Refractive index is a ratio, it does not have a unit.

The refractive index is the factor by which the speed of light is reduced when it passes from a vacuum into the material. This is sometimes called the **absolute** refractive index to distinguish it from the **relative** refractive index between two materials.

> **KEY POINT**
>
> The refractive index between two materials, $_1n_2$ is the ratio of the speed of light in material 1 to the speed of light in material 2.
> $$_1n_2 = c_1 \div c_2 = n_1 \div n_2$$

The refractive index of air is 1.0003, so there is no difference in the refractive index of a material relative to air and relative to a vacuum when working to three significant figures.

The greater the change in speed when light is refracted, the greater the change in direction. At a given angle of incidence, there is a greater change in direction when light passes into glass, $n_g = 1.50$, than when light passes into water, $n_w = 1.33$. The values quoted here are absolute values, but in practice there is little difference between the refractive index of a material relative to a vacuum and that relative to air.

When light is refracted:

* the incident light, refracted light and the normal all lie in the same plane
* **Snell's law** relates the change in direction to the change in speed that takes place.

> **KEY POINT**
>
> Snell's law states that:
> $$\frac{\sin i}{\sin r} = \frac{c_1}{c_2} = {}_1n_2$$

So the sines of the angles are in the same ratio as the speeds in the media when refraction takes place.

Critical angle

AQA A	M1	OCR A	M3
AQA B	M2	OCR B	M1
EDEXCEL B	M1	WJEC	M1
NICCEA	M2		

When light speeds up as it crosses a boundary, the change in direction is away from the normal line. For some angle of incidence, called the **critical angle**, the angle of refraction is 90°.

The diagram below shows what happens to light meeting a glass–air boundary at angles of incidence that are **a** less than, **b** equal to and **c** greater than the critical angle.

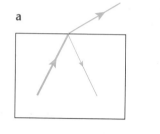

a

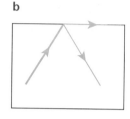

b

c

> The intensity of the internal reflection increases as the angle of incidence increases.

there is a weak reflection   there is a stronger reflection   all light is reflected

These diagrams show that:

- at angles of incidence less than the critical angle, both reflection and refraction take place
- at the critical angle, the angle of refraction is 90°
- at angles of incidence greater than the critical angle, the light is **totally internally reflected**.

> The abbreviation TIR is often used for total internal reflection.

Snell's law can be used to show that the relationship between the critical angle and refractive index is:

> Diamond has a very high refractive index and a low critical angle. Multiple internal reflections make diamonds sparkle.

$$n = 1/\sin C$$
where *C* is the critical angle and *n* is the absolute refractive index.

KEY POINT

Fibre optics

AQA A	M1	OCR A	M3
AQA B	M2	OCR B	M1
EDEXCEL B	M1	WJEC	M1

Total internal reflection is used in prismatic binoculars, car and cycle reflectors and cats' eyes, but the greatest impact has been in the fields of medicine and communications.

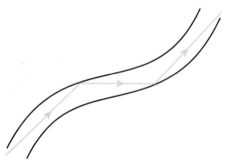

Optical fibres enable light to travel round curves by repeated total internal reflection at the boundaries of the fibre, which is made from glass or plastic. Provided that light hits the boundary at an angle greater than the critical angle, none passes out of the fibre. The diagram opposite shows how light can be made to travel 'round the bend' of a fibre.

In medicine, optical fibres are used in **endoscopy** to look inside a patient. This can be done using a natural body opening or by 'keyhole surgery', which requires a small incision just wide enough to take the fibre. Two bundles of fibres are used in endoscopy; one to transmit light to inside the patient and the other to carry the reflected light to a small television camera.

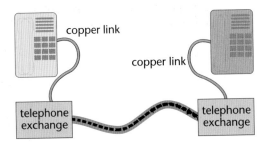

In communications, optical fibres are used to transmit telephone conversations, television and radio programmes in the form of **digital signals**. The signals used can only have the values 1 or 0, where 1 is represented by a pulse of light from a laser diode and 0 is the absence of light. The diagram opposite shows how optical fibres are used in telephone communications.

> Although they are commonly referred to as light, the pulses used are usually in the infra-red region of the electromagnetic spectrum.

The advantages of using optical fibres rather than copper cables in communications are:

- the range of a signal in an optical fibre is much greater than that in a copper cable, so amplification is needed less frequently
- optical fibres do not pick up **noise** from changing magnetic fields.

> The range of a signal in a copper cable is only a few km, compared to up to 100 km in a glass fibre.

However, as the diagram above shows, the signal in an optical fibre is still subject to distortion as it travels along the fibre. This is due to **multipath dispersion**, a process which causes the pulse to become elongated as some parts of the pulse travel further than others, for example when the cable goes round corners. In modern fibres the effect of this is reduced by concentrating the light pulses along the core of the fibre.

Multipath dispersion and its effect on a digital signal

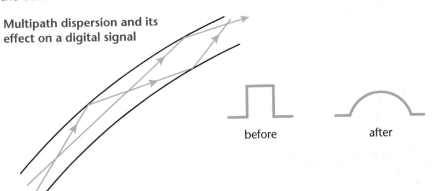

before   after

> A similar effect occurs if 'light' containing more than one wavelength is used, as the different wavelengths travel at different speeds in the fibre.

Transmission of information using digital, rather than analogue signals has two main advantages:

- any noise or distortion can be easily removed in a process called regeneration
- digital signals can carry more information than analogue signals.

Noise affects the amplitude of a signal. In AM (amplitude modulated) transmissions, the information is carried in the form of variations in amplitude of a wave. Any noise cannot be distinguished from the signal and so cannot be removed. When the signal is amplified, the noise is also amplified.

This is not the case with a digital signal, which only has certain allowed values. If the amplitude varies around these values, it can be restored. This is shown in the diagram below.

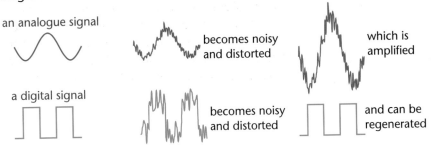

an analogue signal — becomes noisy and distorted — which is amplified

a digital signal — becomes noisy and distorted — and can be regenerated

> The 'allowed values' used in communications are normally 1 and 0, so regeneration is a straightforward task.

Progress check

The table gives the speed of light in different media.

medium	speed of light/m s⁻¹
vacuum	3.00×10^8
ice	2.29×10^8
ethanol	2.21×10^8
quartz	1.94×10^8

1 a Calculate the values of the absolute refractive indexes of ice, ethanol and quartz.

b Calculate the refractive index for light travelling from ice into ethanol.

2 A piece of ice is placed in ethanol.

a In which direction is light travelling for total internal reflection to take place?

b Calculate the value of the critical angle.

3 The critical angle for a certain glass is 39°. Calculate the refractive index of the glass.

3 1.59
b 75°
2 a from ethanol to ice
b 1.04
1 a ice 1.31 ethanol 1.36 quartz 1.55

4.3 More complex wave behaviour

After studying this section you should be able to:

● describe how the diffraction of a wave at a gap depends on the wavelength and size of gap
● explain how wave superposition causes interference patterns and describe the conditions for the patterns to be observable
● explain the formation of a stationary wave

LEARNING SUMMARY

Diffraction

AQA A	A2	NICCEA	M2
AQA B	M2	OCR A	M2, M3
EDEXCEL A	A2	OCR B	M2
EDEXCEL B	M2	WJEC	M1

Our everyday experience is that sound can travel round corners, but light travels in straight lines. The spreading out of waves as they pass through openings or the edges of obstacles is called **diffraction**. All waves can be diffracted, but the effect is more noticeable with long wavelength waves such as sound and radio waves than with short wavelength waves such as light and X-rays.

The diagrams show what happens when water waves pass through gaps of different sizes.

The amount of spreading when the wave has passed through the gap depends on the relative sizes of the gap and the wavelength.

> When answering questions about diffraction, always emphasise the size of the gap compared to the wavelength.

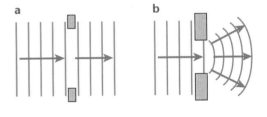

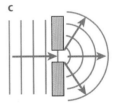

● For a gap that is many wavelengths wide, see diagram **a**, no detectable spreading takes place.
● Some spreading occurs when waves pass through a gap that is several wavelengths wide, see diagram **b**.
● The maximum amount of spreading corresponds to a gap that is the same size as the wavelength as in diagram **c**.

> Diffraction explains why you can 'hear' round corners but you cannot 'see' round corners.

Diagram **a** models what happens when light (wavelength approximately 5×10^{-7} m) passes through a doorway, and diagram **c** models sound (wavelength approximately 1 m) passing through the same doorway.

Diffraction also explains why long wavelength radio broadcasts can be detected in the shadows of hills and buildings, but short wavelength broadcasts cannot. Diffraction is also important in the design of loudspeakers to maximise the spreading of sounds of all wavelengths and in satellite transmissions to minimise the diffraction that occurs.

Superposition and interference of sound

AQA A	A2	NICCEA	M2
AQA B	M2	OCR A	M3
EDEXCEL A	A2	OCR B	M2
EDEXCEL B	M1	WJEC	M1

The **superposition** of waves occurs when two or more waves cross. This is happening all the time, but the effects are so short-lived that we seldom notice them. One of the easiest effects of superposition to observe is **two-source interference**. This can be demonstrated with sound, light, surface water waves and 3 cm electromagnetic waves. The diagram overleaf shows the variation in the loudness of the sound in front of two loudspeakers vibrating in phase.

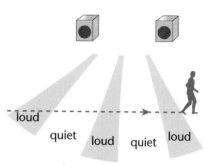

An observer walking along a line parallel to the loudspeakers notices that:

- there are regions where the sound is much louder than from one loudspeaker alone
- there are regions where the sound is barely audible
- in between these, the intensity of the sound varies from loud to quiet.

This is explained using the **principle of superposition**, which describes how two waves can reinforce each other or cancel each other out.

Superposition of sound waves

> The term 'vector sum' here means that the directions are taken into account, so that a positive displacement can be cancelled by a negative one.

> **KEY POINT**
> When two or more waves cross, the resultant displacement is equal to the vector sum of the individual displacements.

This means that displacements in the same direction cause reinforcement, while displacements in opposite directions cause cancellation. The diagram opposite shows how a loud sound and no sound can be produced when two waves of equal amplitude meet at a point.

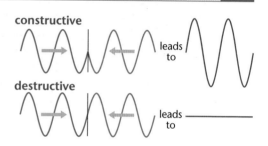

- If the waves are exactly in step (in phase) then **constructive interference** takes place, resulting in a loud sound.
- If the waves are exactly out of step (out of phase) then **destructive interference** takes place, resulting in no sound.

Conditions for interference to be observed

AQA A	A2	NICCEA	M2
AQA B	M2	OCR A	M3
EDEXCEL A	A2	OCR B	M2
EDEXCEL B	M1	WJEC	M1

> The amplitudes do not need to be the same, but they should be comparable for the interference to be observed. If one wave has a much greater amplitude than the other then the effects of cancellation and reinforcement will not be noticeable.

The waves do not need to have the same amplitude for interference to take place. The effect of one wave having a greater amplitude than the other is that the destructive interference is not total. This happens when the observer is closer to one loudspeaker than to the other.

The diagram below represents the waves from two sources vibrating in phase. The solid and broken lines could represent peaks and troughs in the case of water waves or compressions and rarefactions in the case of sound waves.

> Try drawing similar patterns and investigating the effect of changing the wavelength and the separation of the sources.

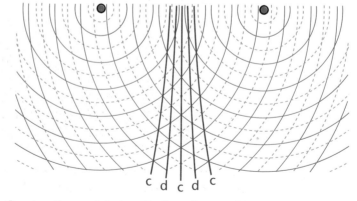

Constructive and destructive interference from two wave sources

Some lines of constructive interference (c) and destructive interference (d) have been marked in.

If the sources in the diagram on page 126 are in phase, then at any point equidistant from both sources the waves arrive in phase, giving constructive interference. This is the case for points along the central c line in this diagram. At all other points in the interference pattern, there is a **path difference**, i.e. the wave from one source has travelled further than that from the other source.

> The conditions can be written as $n\lambda$ path difference for constructive interference and $(2n+1)\lambda/2$ path difference for destructive interference, where n is an integer.

- **Constructive interference** takes place when the path difference is a whole number of wavelengths.
- **Destructive interference** takes place when the path difference is one and a half wavelengths, two and a half wavelengths, etc.

This is shown in the diagram.

the path difference here is half a wavelength, giving destructive interference

the path difference here is one wavelength, giving constructive interference

> Clearly, for an interference pattern to be observed, the sources must not only be coherent but also of the same type.

These conditions for constructive and destructive interference only apply to two sources that are in phase. If the phase of one of the oscillators in the diagram on page 126 is reversed so that the sources are in antiphase, then the conditions for constructive and destructive interference are also reversed. The effect is to shift the interference pattern so that lines of constructive interference (c) become destructive and vice versa. Any value of phase difference between the two sources gives an interference pattern that is stationary, provided that the phase difference is not changing. Two sources with a **fixed phase difference** are said to be **coherent**. Coherent sources must have the same wavelength and frequency.

Interference of light

AQA A	A2	NICCEA	M2
AQA B	M2	OCR A	M3
EDEXCEL A	A2	OCR B	M2
EDEXCEL B	M1	WJEC	M1

Interference patterns are easy to set up with water waves and sound because two sources can be driven from the same oscillator, giving coherence. Because of the way in which light is emitted in random bursts of energy from a source, it is not possible to have two separate sources that are coherent. Instead, diffraction is used to obtain two identical copies of the light from a single source. This is done by illuminating two narrow slits with a lamp that is parallel to the slits so that the same wavefronts arrive at each slit. A suitable arrangement is shown in the diagram below.

> A colour filter can be used between the lamp and the slits to reduce the range of wavelengths interfering, but this also reduces the intensity of the interference pattern.

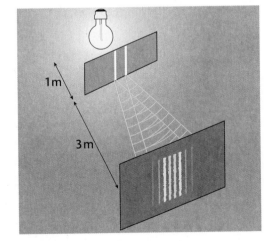

Bright and dark fringes on the screen are due to constructive and destructive interference between the two overlapping beams of light. The separation of the fringes depends on:

- the separation of the slits
- the wavelength of the light
- the distance between the slits and the screen.

This formula enables the wavelength of light to be measured from a simple experiment. The fringe spacing, x, should be obtained by measuring the separation of as many fringes as are visible and dividing by the number of fringes.

The separation of the fringes is related to the other variables by the formula:

$$\text{fringe spacing} = \frac{\text{wavelength} \times \text{distance from slits to screen}}{\text{slit separation}}$$

$$x = \lambda D/a$$

where x is the distance between adjacent bright (or dark) fringes
λ is the wavelength of the light
D is the distance between the slits and the screen
a is the distance between the slits.

As the wavelength of light is very small, the slits need to be close together and separated from the screen by a large distance for the interference fringes to be seen.

Stationary (standing) waves

AQA A	A2	NICCEA	M2
AQA B	M2	OCR A	M3
EDEXCEL A	A2	OCR B	M2
EDEXCEL B	M1	WJEC	M1

A progressive wave is one whose profile moves through space, whereas a **stationary** wave has a static profile. The waves on the vibrating strings and in the vibrating air columns of musical instruments are stationary waves.

Stationary waves are caused by the superposition of two waves of the same wavelength travelling in opposite directions. They often arise when a wave is reflected at a boundary, but the waves can come from two separate coherent sources. The diagram shows the vibrations of a stationary wave on a string.

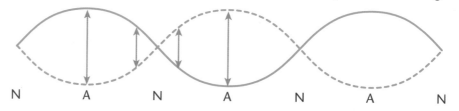

N   A   N   A   N   A   N

All points on a progressive wave vibrate with the same amplitude, but each point is only in phase with points precisely n wavelengths away.

Stationary waves differ from progressive waves in a number of respects:
- there is no flow of energy along a stationary wave, although stationary waves often radiate energy
- within each loop of a stationary wave, all particles vibrate in phase and exactly out of phase (180° phase difference) with the particles in adjacent loops
- the amplitude of vibration varies with position in the loop
- there are **nodes** (points marked N in the diagram), where the displacement is always zero and **antinodes** (marked A in the diagram) which vibrate with the maximum amplitude.

A convenient way of measuring the wavelength of a progressive wave is to use it to set up a stationary wave, for example by superimposing the wave on its reflection from a barrier, and measuring the distance between a number of successive nodes or antinodes.

The wavelength of a stationary wave is twice the distance between two adjacent nodes or antinodes. The diagram opposite shows how a stationary wave (shown with a solid line) is formed from two progressive waves (shown with broken lines) travelling in opposite directions.

In (a) the waves interfere constructively, so each point on the wave has its maximum displacement.

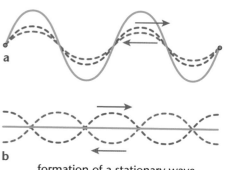

a

b

formation of a stationary wave

In (b) each progressive wave has moved one quarter of a wavelength; the interference is now destructive, resulting in no displacement at all points on the wave.

Progress check

1 When light passes through a slit which is 2 mm wide, it produces a narrow beam. Explain why there is no observable diffraction.

2 For an observable interference pattern between two sources, the sources must be **coherent**.
  a What is meant by two coherent sources?
  b What TWO other conditions should be met for an observable interference pattern?

3 In a two-slit interference experiment using light, the slits are separated by a distance of 1.2 mm. The distance from the slits to the screen is 3.0 m and the separation of the bright fringes is 1.4 mm.
  Calculate the wavelength of the light.

3 5.6×10^{-7} m
  b They should be the same type of wave and similar in amplitude.
2 a Coherent means they have a fixed phase relationship.
1 The width of the slit is very large compared to the wavelength of light.

4.4 Wave properties across the spectrum

After studying this section you should be able to:

- describe the main features of the electromagnetic spectrum
- explain how the photoelectric effect gives evidence for the particulate nature of electromagnetic radiation
- explain line spectra and relate the spectral lines to energy level transitions

The spectrum

AQA B	M2	OCR A	M2
EDEXCEL A	A2	OCR B	M1, M2
EDEXCEL B	M1	WJEC	M2
NICCEA	M2		

The speed of all electromagnetic waves in a vacuum is 2.999×10^8 m s^{-1}. The value 3.00×10^8 m s^{-1} is normally used for both a vacuum and air.

The different changes of speed when light consisting of a range of wavelengths is refracted is responsible for dispersion, the splitting of light into a spectrum.

Microwaves are given here as a separate part of the spectrum, although they can be considered as short-wavelength radio waves.

The electromagnetic spectrum consists of a whole family of waves, with some similarities and some differences in their behaviour.

Similar properties include:

- they consist of electric and magnetic fields oscillating at right angles to each other
- they travel at the same speed in a vacuum
- they are transverse waves, so they can be polarised
- they show the same pattern of behaviour in reflection, refraction, interference and diffraction.

The differences include:

- the shorter wavelength waves undergo a greater change of speed when being refracted
- wave-like properties such as diffraction and interference are more readily observable in the behaviour of the longer wavelength waves.

The diagram shows the range of wavelengths and frequencies of the waves that make up the spectrum.

frequency / Hz	10^{20}	10^{17}	10^{14}	10^{11}	10^8	10^5

gamma rays   ultraviolet   infra-red   radio waves

X-rays   light   microwaves

wavelength / m	10^{-12}	10^{-9}	10^{-6}	10^{-3}	1	10^3

In general, the name given to an electromagnetic wave depends on its wavelength or frequency. Notable exceptions to this are X-rays and gamma rays, whose ranges of wavelength and frequency overlap. The difference here is in the origin of the waves: an X-ray is emitted when a high-speed electron is suddenly brought to rest and a gamma ray is emitted from an excited nucleus, usually along with alpha or beta emission.

The table shows typical wavelengths, the origins of the waves and the uses of the main parts of the electromagnetic spectrum.

name of radiation	typical wavelength/m	how produced	used for
gamma	1×10^{-12}	an excited nucleus releasing energy in radioactive decay	tracing the flow of fluids and treating cancer
X-rays	1×10^{-10}	high-speed electrons being stopped by a target	seeing inside the body
ultraviolet	1×10^{-8}	very hot objects and passing electricity through gases	suntan and security marking

light	1×10^{-6}	hot objects and passing electricity through gases	vision and photography
infra-red	1×10^{-5}	warm and hot objects	heating and cooking
microwave	1×10^{-1}	microwave diode or oscillating electrons in an aerial	communications and cooking
radio	1×10^{2}	oscillating electrons in an aerial	communications

All objects give out infra-red radiation, no matter what the temperature. The hotter the object, the more power is emitted and the greater the range of wavelengths.

Waves or particles

AQA A	M1	NICCEA	M2
AQA B	A2	OCR A	M2
EDEXCEL A	A2	OCR B	M2
EDEXCEL B	M1, M2	WJEC	M2

The **threshold wavelength**, λ_0, is the wavelength of the waves that have the threshold frequency.
$$\lambda_0 = c \div f_0$$

The photoelectric effect can be demonstrated using a zinc plate connected to a gold leaf electroscope. An ultraviolet lamp discharges a negatively-charged plate but has no effect on a positively-charged plate.

The word quantum refers to the smallest amount of a quantity that can exist. A quantum of electromagnetic radiation is the smallest amount of energy of that frequency.

The **photoelectric effect** provides evidence that electromagnetic waves have a particle-like behaviour which is more pronounced at the short-wavelength end of the spectrum. In the photoelectric effect electrons are emitted from a metal surface when it absorbs electromagnetic radiation.

The results of photoelectricity experiments show that:

- there is no emission of electrons below a certain frequency, called the **threshold frequency**, f_0, which is different for different metals
- above this frequency, electrons are emitted with a range of kinetic energies up to a maximum, $(\frac{1}{2}mv^2)_{max}$
- increasing the frequency of the radiation causes an increase in the maximum kinetic energy of the emitted electrons, but has no effect on the photoelectric current, i.e. the rate of emission of electrons
- increasing the intensity of the radiation has no effect if the frequency is below the threshold frequency; for frequencies above the threshold it causes an increase in the photoelectric current, so the electrons are emitted at a greater rate.

The wave model cannot explain this behaviour; if electromagnetic radiation is a continuous stream of energy then radiation of all frequencies should cause photoelectric emission, it should only be a matter of time for an electron to absorb enough energy to be able to escape from the attractive forces of the positive ions in the metal.

The explanation for the photoelectric effect relies on the concept of a **photon**, a quantum or packet of energy. We picture electromagnetic radiation as short bursts of energy, the energy of a photon depending on its frequency.

A lamp emits random bursts of energy. Each burst is a photon, a quantum of radiation.

> **KEY POINT**
>
> The relationship between the energy, E of a photon, or quantum of electromagnetic radiation, and its frequency, f, is:
> $$E = hf$$
> where h is Planck's constant and has the value 6.63×10^{-34} J s.

The energy of a photon can be measured in either joules or **electronvolts**. The electronvolt is a much smaller unit than the joule.

The conversion factor for changing energies in eV to energies in joules is 1.60×10^{-19} J eV^{-1}

> **KEY POINT**
>
> One electron volt (1 eV) is the energy transfer when an electron moves through a potential difference of 1 volt.
> $$1 \text{ eV} = 1.60 \times 10^{-19} \text{ J}$$

Einstein's explanation of photoelectric emission is:

The work function is the **minimum** energy needed to liberate an electron from a metal. Some electrons need more than this amount of energy.

- an electron needs to absorb a minimum amount of energy to escape from a metal. This minimum amount of energy is a property of the metal and is called the **work function**, ϕ
- if the photons of the incident radiation have energy hf less than ϕ then there is no emission of electrons
- emission becomes just possible when $hf = \phi$
- for photons with energy greater than ϕ, the electrons emitted have a range of energies, those with the maximum energy being the ones that needed the minimum energy to escape

Radiation below the threshold frequency, f_0, no matter how intense, does not cause any emission of electrons.

- increasing the intensity of the radiation increases the number of photons incident each second. This causes a greater emission of electrons, but does not affect their maximum kinetic energy

> **KEY POINT**
>
> Einstein's photoelectric equation relates the maximum kinetic energy of the emitted electrons to the work function and the energy of each photon:
>
> $$hf = \phi + (\tfrac{1}{2}mv^2)_{max}$$

At the threshold frequency, the minimum frequency that can cause emission from a given metal, $(\tfrac{1}{2}mv^2)_{max}$ is zero and so the equation becomes $hf_0 = \phi$.

Particles or waves

AQA A	M1	NICCEA	M2
AQA B	A2	OCR A	M2
EDEXCEL A	A2	OCR B	M2
EDEXCEL B	M1, M2	WJEC	M2

If waves can show particle-like behaviour in photoelectric emission, can particles also behave as waves? Snooker balls bounce off cushions in the same way that light bounces off a mirror, so reflection is not a test for wave-like or particle-like behaviour. Diffraction and interference are properties unique to waves, so particles can be said to have a wave-like behaviour if they show these properties.

All particles have an associated wavelength called the **de Broglie** wavelength.

Try calculating the de Broglie (pronounced de Broy) wavelength of a moving snooker ball. Is it possible for the ball to show wave-like behaviour?

> **KEY POINT**
>
> The wavelength, λ, of a particle is related to its momentum, p, by the de Broglie equation:
>
> $$\lambda = h/p = h/mv$$
>
> where h is the Planck constant.

An electron that has been accelerated through a potential difference of a few hundred volts has a wavelength similar to that of X-rays and gamma rays.

The de Broglie wavelength of such an electron is of the order of 1×10^{-10} m.

This wavelength is also similar to the spacing of the atoms in crystalline materials, so these materials provide suitable sized gaps to cause diffraction.

Diffraction patterns formed by a beam of electrons after passing through thin foil or graphite show a set of bright and dark rings on photographic film, similar to those formed by X-ray diffraction.

You may have seen this pattern formed on a fluorescent screen in a vacuum tube.

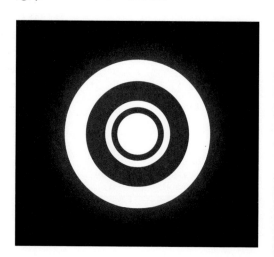

a diffraction pattern formed by passing a beam of electrons through graphite

Electrons can also be made to interfere when two coherent beams overlap. They produce an interference pattern similar to that of light, but on a much smaller scale.

Other particles such as protons and neutrons also show wave-like behaviour.

Particles and waves are the models that we use to describe and explain physical phenomena. It is not surprising that the real world does not fit neatly into our models.

There are two separate models of how matter behaves. The particle model explains such phenomena as ionisation and photoelectricity, while the wave model explains interference and diffraction. It is not appropriate to classify matter as waves or particles as photons and electrons can fit either model, depending on the circumstances.

Line spectra

AQA A	M1	NICCEA	M2
AQA B	M2	OCR B	M2
EDEXCEL A	A2	WJEC	M2
EDEXCEL B	M1, M2		

Gamma rays are the exception because they are a result of the nucleus losing energy.

With the exception of gamma rays, the emission of electromagnetic radiation is associated with electrons losing energy. A hot solid can radiate the whole range of wavelengths through the infra-red and part of the visible spectrum, the extent of the range depending on its temperature.

When an electric current is passed through an ionised gas, only a small number of wavelengths are emitted. The wavelengths are characteristic of the gas used and are called a **line spectrum**.

The diagram below shows some of the lines present in the hydrogen spectrum and their wavelengths.

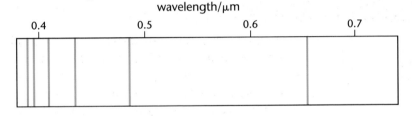

Not all of these lines lie in the visible spectrum, which extends from 0.4 μm to 0.65 μm.

The existence of line spectra provides evidence that the electrons in orbit around a nucleus can only have certain values of energy, the values being characteristic of an atom. Energy can only be emitted or absorbed in amounts that correspond to the differences between these allowed values.

An **energy level diagram** shows the amounts of energy that an electron can have. The diagram below is an energy level diagram for hydrogen. Note that on an energy level diagram:

- the energies are measured relative to a zero that represents the energy of an electron at rest outside the atom, i.e. one that is just free
- an orbiting electron has less energy than a free electron, so it has negative energy relative to the zero
- an electron with the minimum possible energy is in the **ground state**; higher energy levels are called **excited states**.

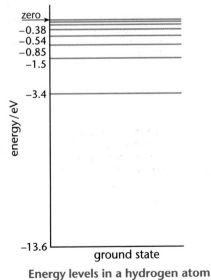

Energy levels in a hydrogen atom

Movement of an electron to the ground state results in the emission of a photon with an energy in excess of 10 eV. Photons with this amount of energy give spectral lines in the ultraviolet region of the spectrum.

The lines of the emission spectrum often appear black in the absorption spectrum. The photons of these energies are absorbed as the electrons move to more excited states, and then released when the electrons lose energy. The emitted photons are radiated in all directions, so very little energy is detected in any one direction.

Energy is emitted in the form of a photon when an electron moves from an excited state to a lower energy level. The energy of the photon is equal to the difference in the values of the energy levels. For example, an electron moving from an energy level with –0.38 eV of energy to one with –0.85 eV loses energy equal to (–0.38 – –0.85) eV = 0.47 eV. This corresponds to emitting a photon of frequency 1.13×10^{14} Hz, which lies in the infra-red part of the electromagnetic spectrum.

Electrons can gain energy by absorbing photons. As with emission, the only photons that can be absorbed are those that correspond to allowed movements, or transitions, of electrons. An **absorption spectrum** is produced by shining white light through a sample of a gaseous element. The spectrum that emerges is the full spectrum with the element's emission spectrum missing or of low intensity. This is due to the electrons absorbing photons of just the right energy to allow them to move to a more excited state.

> **KEY POINT**
>
> When an electron moves from an energy level E_1 to a lower energy level E_2, the energy of the photon emitted is given by
> $$hf = E_1 - E_2$$

Progress check

1 A photon of green light has a wavelength of 5.0×10^{-7} m.
   a Calculate the frequency of the light.
   b Calculate the energy of the photon:
      i   in J
      ii in eV.

2 Calculate the de Broglie wavelength of an electron, $m_e = 9.1 \times 10^{-31}$ kg, travelling at a speed of 3.0×10^7 m s^{-1}.

3 An electron in a hydrogen atom undergoes a transition from an energy level of –0.54 eV to one of –3.40 eV.
   a Calculate the frequency of the photon emitted.
   b In what part of the electromagnetic spectrum is the emitted radiation?

3 a 6.90×10^{14} Hz   b visible (light)
2 2.43×10^{-11} m
1 a 6.0×10^{14} Hz   b i 3.98×10^{-19} J   ii 2.49 eV

Sample question and model answer

In the photoelectric effect, electrons are emitted from a metal surface when it is irradiated with electromagnetic radiation. The graph shows the variation of the maximum photoelectron kinetic energy with the frequency of the radiation incident on the emitting surface.

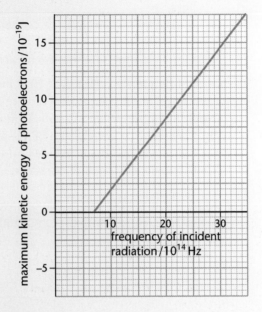

The equation needs to be written in the form $y = mx + c$ to determine which aspect of the graph determines the Planck constant, h.
The graph is a plot of $(\frac{1}{2}mv^2)_{max}$ against f, so rearrange the equation with $(\frac{1}{2}mv^2)_{max}$ on the left-hand side and f on the right-hand side.

(a) Use the data from the graph to calculate the Planck constant. [3]

The photoelectric equation is
$$hf = \theta + (\frac{1}{2}mv^2)_{max}$$
This can be written as
$$(\frac{1}{2}mv^2)_{max} = hf - \theta$$ 1 mark

It is important to remember here that a graph is a relationship between two sets of numbers, so the gradient does not have a unit. The gradient of this graph gives the numerical value of the Planck constant; you need to add the correct unit to this numerical value in your final answer.

The gradient of a graph of $(\frac{1}{2}mv^2)_{max}$ against f represents the Planck constant. 1 mark

The gradient of the graph $= 17 \times 10^{-19} \div (34 - 7.0) \times 10^{14} = 6.30 \times 10^{-34}$.
This gives the value of the Planck constant as 6.30×10^{-34} Js. 1 mark

(b) Determine the minimum energy required to remove an electron from the target metal. [2]

The minimum energy required to remove an electron is hf_0, where f_0 is the threshold frequency. 1 mark
$$hf_0 = 6.30 \times 10^{-34} \text{ Js} \times 7.5 \times 10^{14} \text{ Hz} = 4.73 \times 10^{-19} \text{ J}$$ 1 mark

This value is obtained by using the unrounded value of h from part a. If you use the rounded value, you should obtain an answer of 4.82×10^{-19} J. Either answer gains full marks.

(c) Explain how the photoelectric effect produces evidence which illustrates the particulate nature of light. [3]

There is no emission below the threshold frequency; this shows that energy arrives in packets rather than a continuous stream. 1 mark
The maximum kinetic energy of the emitted electrons depends on the frequency of the radiation. This shows that the energy of each packet of radiation depends on the frequency. 1 mark
Increasing the intensity of the radiation does not affect the maximum kinetic energy of the emitted electrons, only the rate of emission. This shows that intensity is related to the number of packets of energy incident each second. 1 mark

Since the question asks how the evidence supports the particulate nature of light, it is important to explain how each piece of evidence quoted supports this theory, rather than just quoting the evidence.

AEB AS-A/PHYS/7 0635J Jan 1999

Practice examination questions

1 (a) Distinguish between a **transverse** and a **longitudinal** wave and give one example of each. [4]

(b) Light can be **polarised** when it is reflected at a surface.
(i) Describe the difference between polarised light and unpolarised light. [2]
(ii) Explain why sound cannot be polarised. [1]

2 (a) State the laws that describe the refraction of light. [2]

(b) Light of frequency 6.00×10^{14} Hz is incident on an air–water boundary at an angle of 63°. The speed of light in air is 3.00×10^8 m s^{-1} and the speed of light in water is 2.25×10^8 m s^{-1}.
(i) Calculate the wavelength of the light in air. [3]
(ii) Calculate the refractive index for light passing from air into water. [2]
(iii) Calculate the angle of refraction in the water. [2]

3 The diagram shows how the profile of a wave on a rope changes. In a period of 2.50×10^{-4} s it moves from position 1 to position 2.

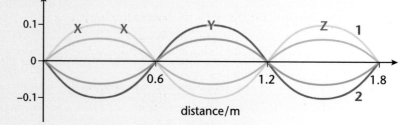

(a) (i) Write down the wavelength of the wave. [1]
(ii) Calculate the frequency of the wave. [2]
(iii) Calculate the speed of the wave along the rope. [3]

(b) (i) How can you tell from the diagram that the wave is a stationary wave? [1]
(ii) Suggest how the stationary wave is formed in this example. [2]

(c) What is the phase difference between:
(i) the two points marked X on the diagram? [1]
(ii) the points marked Y and Z on the diagram? [1]

Give your answers in degrees.

Practice examination questions (continued)

4 The diagram shows two identical loudspeakers, A and B, placed 0.75 m apart. Each loudspeaker emits sound of frequency 2000 Hz.
Point C is on a line midway between the speakers and 5.0 m away from the line joining the speakers. A listener at C hears a maximum intensity of sound. If the listener then moves from C to E or D, the sound intensity heard decreases to a minimum. Further movement in the same direction results in the repeated increase and decrease in the sound intensity.

speed of sound in air = 330 m s^{-1}

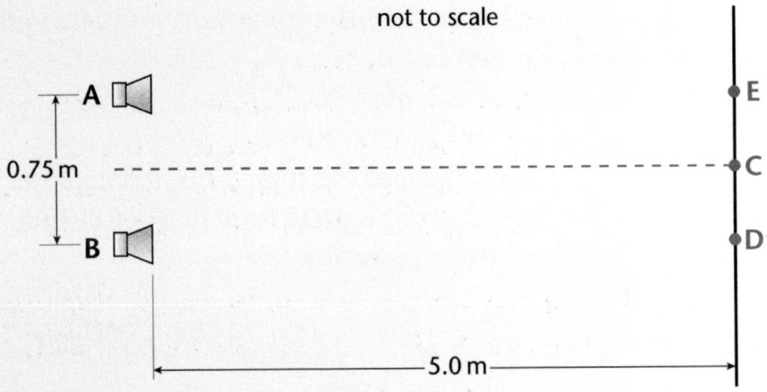

not to scale
A
B
0.75 m
E
C
D
5.0 m

(a) Explain why the sound intensity is:
  (i) a maximum at C,
  (ii) a minimum at D or E. [4]

(b) Calculate:
  (i) the wavelength of the sound,
  (ii) the distance CE. [4]

NEAB Particles and Waves (PH02), June 1999, Q 5

5 The diagram shows an optical fibre made of a plastic-coated glass.
The refractive index of the glass is 1.58 and that of the plastic is 1.24.

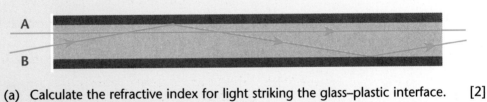

A
B

(a) Calculate the refractive index for light striking the glass–plastic interface. [2]

(b) Calculate the value of the critical angle at this interface. [2]

(c) The diagram shows two paths of light through the fibre.
Explain how this can affect a pulse of light as it passes along a fibre. [2]

Practice examination questions *(continued)*

6 When electromagnetic radiation is incident on a clean surface of copper, *photoelectric emission* may take place. No emission occurs unless the photon energy is greater than or equal to the *work function*.

(a) Explain the meaning of the terms:

    (i) photoelectric emission   [2]

    (ii) photon   [2]

    (iii) work function   [2]

(b) The work function of copper is 5.05 eV.

    (i) Calculate the minimum frequency of radiation that causes photoelectric emission.

        $e = 1.60 \times 10^{-19}$ C

        $h = 6.63 \times 10^{-34}$ Js   [2]

    (ii) Calculate the maximum kinetic energy of the emitted electrons when electromagnetic radiation of frequency 1.80×10^{15} Hz is incident on the copper surface.   [3]

Practice examination answers

Chapter 1   Force and motion 1

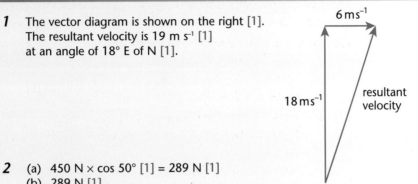

1   The vector diagram is shown on the right [1].
     The resultant velocity is 19 m s⁻¹ [1]
     at an angle of 18° E of N [1].

2   (a) 450 N × cos 50° [1] = 289 N [1]
     (b) 289 N [1]
     (c) The light fitting is in equilibrium [1] so the sum of the forces on it is zero [1].
     (d) The vector diagram is shown on the right [1].
          S = 345 N [1].

3   (a) Moment = force × distance to pivot = 60 N × 0.45 m [1]
                = 27 N m [1].
     (b) F = 27 N m ÷ 0.06 m [1] = 450 N [1].
     (c) A bigger moment is caused for the same applied force [1]. So the force acting on the
          branch is greater [1].

4   (a) Horizontal component = 2893 N [1].
          Vertical component = 3447 N [1].
     (b) pressure = normal force ÷ area [1] = 3447 N ÷ 0.09 m² [1] = 38300 Pa [1].
     (c) To reduce the pressure on the ground [1] so that the girder does not penetrate the
          surface [1].

5   (a) Volume of water = 10.5 m × 4.2 m × 3.5 m = 154 m³ [1]
          Mass = volume × density = 1.54 × 10⁵ kg [1]
          Weight = mass × g = 1.54 × 10⁶ N [1].
     (b) Pressure = normal force ÷ area [1] = 1.54 × 10⁶ N ÷ 4.41 × 10¹ m² [1] = 3.5 × 10⁴ Pa [1].
     (c) Liquid pressure acts equally in all directions [1] so at the bottom of the tank the pressure
          on the sides is equal to that on the base [1].
     (d) It is the same [1] the pressure depends on the depth only [1]

6   (a) A ductile material undergoes plastic deformation [1] but a brittle material breaks [1].
     (b) (i)   340 MPa [1]
          (ii)   up to 300 MPa [1]
          (iii) E = stress ÷ strain = 3.00 × 10⁸ Pa ÷ 1.50 × 10⁻³ [1] = 2.00 × 10¹¹ Pa [1].
          (iv)   The curve would have the same shape [1]; the values on the axes would be the
                 same as the stress-strain graph is the same for all samples of the material [1].

7   (a) Acceleration = change in velocity ÷ time [1] = 24 m s⁻¹ ÷ 80 s [1] = 0.30 m s⁻² [1].
     (b) Between 230 and 290 s [1].
     (c) (i)   20 s [1]
          (ii)   120 s [1].
     (d) 960 m + 1680 m + 720 m + 900 m + 1500 m + 900 m = 6660 m [2] (for all correct).
     (e) 3360 m – 3300 m = 60 m [1].

8
     (a) Use $s = ut + \frac{1}{2}at^2$ [1] $t = \sqrt{(2 \times 0.39 \text{ m} \div 10 \text{ m s}^{-2})}$ [1] = 0.28 s [1].
     (b) $s = ut$ [1] = 35 m s⁻¹ × 0.28 s = 9.8 m [1].
     (c) $u = s \div t$ [1] = 15 m ÷ 0.28 s = 54 m s⁻¹ [1].

9
     (a) $t = \Delta v \div a$ [1] = 14 m s⁻¹ ÷ 2 m s⁻² = 7 s [1].
     (b) distance = average speed × time [1] = 21 m s⁻¹ × 7 s = 147 m [1].

10
     (a) Use $v^2 = u^2 + 2as$ [1] u^2 = – 2 × – 10 m s⁻² × 3.5 m = 70 m²s⁻² [1] u = 8.4 m s⁻¹ [1].
     (b) Time travelling upwards, t = distance ÷ average speed = 3.5 m ÷ 4.2 m s⁻¹ = 0.83 s [1]
          Time in air = 1.66 s [1].

Chapter 2 Force and motion 2

1 (a) Use $v^2 = u^2 + 2as$ [1] $a = v^2 \div 2s = (60 \text{ m s}^{-1})^2 \div 2 \times 1500 \text{ m}$ [1] $= 1.2 \text{ m s}^{-2}$ [1].

(b) $F = ma$ [1] $= 70\,000 \text{ kg} \times 1.2 \text{ m s}^{-2}$ [1] $= 84 \text{ kN}$ [1].

(c) As the aircraft gains speed the size of the resistive forces increases [1]. This causes the resultant force to decrease [1].

2 (a) (i) The pull of the trailer on the car [1]

(ii) 150 N backwards [1].

(b) (i) 130 N forwards [1]

(ii) $a = F \div m$ [1] $= 130 \text{ N} \div 190 \text{ kg}$ [1] $= 0.68 \text{ m s}^{-2}$ [1].

(c) (i) 0 [1]

(ii) 150 N backwards [1].

3 (a) (i) $40 \text{ kg} \times g = 400 \text{ N}$ [1]

(ii) The child pulls the Earth [1].

(b) (i) $400 \text{ N} \times \cos 66°$ [1] $= 163 \text{ N}$ [1]

(ii) $a = F/m$ [1] $= 73 \text{ N} \div 40 \text{ kg}$ [1] $= 1.8 \text{ m s}^{-2}$ [1].

(c) Use $v^2 = u^2 + 2as$ [1] $v^2 = 2 \times 1.8 \text{ m s}^{-2} \times 5.5 \text{ m} = 19.8 \text{ m}^2 \text{ s}^{-2}$ [1] $v = 4.4 \text{ m s}^{-1}$ [1].

4 (a) Momentum before and after collision $= 0.45 \text{ kg} \times 0.60 \text{ m s}^{-1} - 0.30 \text{ kg} \times 0.42 \text{ m s}^{-1}$ [1] $= 0.144 \text{ kg m s}^{-1}$ [1]. Combined speed after collision $= 0.144 \text{ kg m s}^{-1} \div 0.75 \text{ kg} = 0.19 \text{ m s}^{-1}$ [1].

(b) Momentum of lighter vehicle before collision $= (-) 0.45 \text{ kg} \times 0.60 \text{ m s}^{-1} = (-) 0.27 \text{ kg m s}^{-1}$ [1]

Speed $= 0.27 \text{ kg m s}^{-1} \div 0.30 \text{ kg} = 0.90 \text{ m s}^{-1}$ [1].

5 (a) Total momentum before the neutron is absorbed $= 1.7 \times 10^{-27} \text{ kg} \times 1.4 \times 10^7 \text{ m s}^{-1} = 2.38 \times 10^{-20} \text{ kg m s}^{-1}$ [1]. Total momentum after the neutron is absorbed $= 2.38 \times 10^{-20} \text{ kg m s}^{-1}$ [1]. Velocity of nucleus $= p \div m = 2.38 \times 10^{-20} \text{ kg m s}^{-1} \div 4.017 \times 10^{-25} \text{ kg} = 5.9 \times 10^4 \text{ m s}^{-1}$ [1].

(b) The velocity of the nucleus would be greater [1]. The neutron would rebound and have a greater change in momentum [1] so the change in momentum of the nucleus would also be greater [1].

(c) $E_k = \frac{1}{2}mv^2$ [1] $= \frac{1}{2} \times 1.7 \times 10^{-27} \text{ kg} \times (1.4 \times 10^7 \text{ m s}^{-1})^2$ [1] $= 1.67 \times 10^{-13} \text{ J}$ [1].

6 (a) The astronaut gains momentum in the opposite direction to the push on the spacecraft [1]. The spacecraft gains the same amount of momentum, in the direction of the astronaut's push [1].

(b) The change in momentum of the gas [1] is balanced by an equal and opposite change in momentum of the astronaut [1].

7 (a) Change in momentum $= 0.080 \text{ kg} \times 32 \text{ m s}^{-1} - 0.080 \text{ kg} \times -22 \text{ m s}^{-1}$ [1] $= 4.32 \text{ kg m s}^{-1}$ [1].

(b) $F = \Delta p \div t$ [1] $= 4.32 \text{ kg m s}^{-1} \div 0.15 \text{ s}$ [1] $= 28.8 \text{ N}$ [1].

(c) Momentum is conserved [1]. The wall and the Earth gain equal momentum in the opposite direction [1].

8 (a) $W = F \times s$ [1] $= 30\,000 \text{ N} \times 18 \text{ m}$ [1] $= 540 \text{ kJ}$ [1].

(b) 540 kJ [1]

(c) $P = W \div t = 540 \text{ kJ} \div 24 \text{ s}$ [1] $= 22.5 \text{ kW}$ [1].

(d) $P_{in} = P_{out} \div \text{efficiency}$ [1] $= 22.5 \text{ kW} \div 0.45 = 50 \text{ kW}$ [1].

9 (a) $\Delta E_p = mg\Delta h$ [1] $= 2.0 \times 10^5 \text{ kg} \times 10 \text{ N kg}^{-1} \times 215 \text{ m}$ [1] $= 4.30 \times 10^8 \text{ J}$ [1].

(b) $E_k = \frac{1}{2}mv^2 = 4.30 \times 10^8 \text{ J}$ [1] $v = \sqrt{(2 \times 4.30 \times 10^8 \text{ J} \div 2.0 \times 10^5 \text{ kg})}$ [1] $= 65.6 \text{ m s}^{-1}$ [1].

(c) Kinetic energy of water leaving turbines each second $= \frac{1}{2}mv^2 = \frac{1}{2} \times 2.0 \times 10^5 \text{ kg} \times (8 \text{ m s}^{-1})^2 = 6.4 \times 10^6 \text{ J}$ [1]. Maximum power input $= 4.30 \times 10^8 \text{ W} - 6.4 \times 10^6 \text{ W} = 4.24 \times 10^8 \text{ W}$ [1].

(d) Efficiency $=$ useful power out $\div$ power in $= 2.5 \times 10^8 \text{ W} \div 4.24 \times 10^8 \text{ W}$ [1] $= 0.59$ [1].

10 (a) Energy $= \frac{1}{2}Fx = \frac{1}{2} \times 25 \text{ N} \times 0.15 \text{ m}$ [1] $= 1.875 \text{ J}$ [1].

(b) (i) $\frac{1}{2}mv^2 = 1.875 \text{ J}$ [1] $v^2 = 2 \times 1.875 \div 0.020 \text{ kg} = 187.5 \text{ m}^2 \text{ s}^{-2}$ [1] $v = 13.7 \text{ m s}^{-1}$ [1]

(ii) $mg\Delta h = 1.875 \text{ J}$ [1] $\Delta h = 1.875 \text{ J} \div 0.20 \text{ N}$ [1] $= 9.4 \text{ m}$ [1].

Chapter 3 Electricity

1 (a) $R = \rho l/A$ [1].
  (b) Resistivity is a property of a material [1]; the resistance of an object also depends on its dimensions [1].
  (c) Cross-sectional area, $A = 2.0 \times 10^{-5}$ m² [1] $R = 3.00 \times 10^{-5}$ Ω m × 1.0×10^{-2} m ÷ 2.0×10^{-5} m² [1] = 1.5×10^{-2} Ω [1].

2 (a) $1/R = 1/10$ Ω + $1/10$ Ω [1] $R = 5$ Ω [1].
  (b) $I = V/R$ [1] = 6.0 V ÷ 20 Ω [1] = 0.30 A [1].
  (c) 0.30 A ÷ 2 [1] = 0.15 A [1].
  (d) $V = IR$ = 0.15 A × 10 Ω [1] = 1.5 V [1].

3 (a) (i) The graph should show the resistance decreasing with increasing temperature [1] in a non-linear way [1]
    (ii) The resistance of a metallic conductor increases with increasing temperature [1] linearly [1]
    (iii) Ohm's law only applies to metallic conductors [1] at constant temperature [1].
  (b) (i) Current in circuit = 9.0 V ÷ 600 Ω = 1.5×10^{-2} A [1]. Thermistor p.d. = 1.5×10^{-2} A × 500 Ω [1] = 7.5 V [1]
    (ii) The p.d. across the thermistor decreases [1] as it has a smaller proportion of the circuit resistance [1]
    (iii) Current in circuit = 9.0 V ÷ 175 Ω = 5.1×10^{-2} A [1]. Resistor p.d. = 5.1×10^{-2} A × 100 Ω [1] = 5.1 V [1]
    (iv) Current in circuit = 3.9 V ÷ 300 Ω = 0.013 A [1]. Resistor p.d. = 5.1 V [1]. Resistor value = 5.1 V ÷ 0.013 A = 392 Ω [1].

4 (a) The current in the cell and the digital voltmeter is negligible [1] so there is no voltage drop due to internal resistance [1]. With the moving coil voltmeter there is a significant current in the cell; the voltage drop is the p.d. across the internal resistance [1].
  (b) $R = V/I$ [1] = 0.05 V ÷ 1.55×10^{-3} A [1] = 32.3 Ω [1].

5 (a) Current in each lamp, $I = P/V$ = 6 W ÷ 12 V = 0.50 A. Current in battery = 4 × 0.50 A = 2.0 A [1].
  (b) $\Delta q = I\Delta t$ [1] = 2.0 A × 60 s [1] = 120 C [1].
  (c) $R = V/I$ [1] = 12 V ÷ 2.0 A [1] = 6.0 Ω [1].

6 (a) (i) $I^2 = P/R$ [1] $I = \sqrt{(10\ \text{W} ÷ 4.7\ \Omega)}$ [1] = 1.46 A [1]
    (ii) $V = IR$ = 1.46 A × 4.7 Ω [1] = 6.9 V [1]
    (iii) $I = AR/\rho$ [1] = 2.0×10^{-7} m² × 4.7 Ω ÷ 5.0×10^{-7} Ω m [1] = 1.88 m [1].
  (b) A smaller length of material is needed [1] for the same resistance [1].

7 (a) Potential divider [1].
  (b) Apply a varying p.d. to component X [1].
  (c) Circuit B varies the current and only allows investigation of the characteristics when the diode is conducting [1]. Circuit A allows the diode to be investigated when it is not conducting and when it is conducting [1].
  (d) (i) The completed table is:

resistance /Ω	10.8	3.55	1.73	1.15	0.80

      [Two marks for all correct.]
    (ii) The graph is shown on the right.
      Marks are awarded for:
      Scales and labelling of axes [1]
      Correct plotting [2]
      Drawing a smooth curve [1]
    (iii) 0.785 V [1]
    (iv) The diode only conducts over a narrow range of voltages [1].

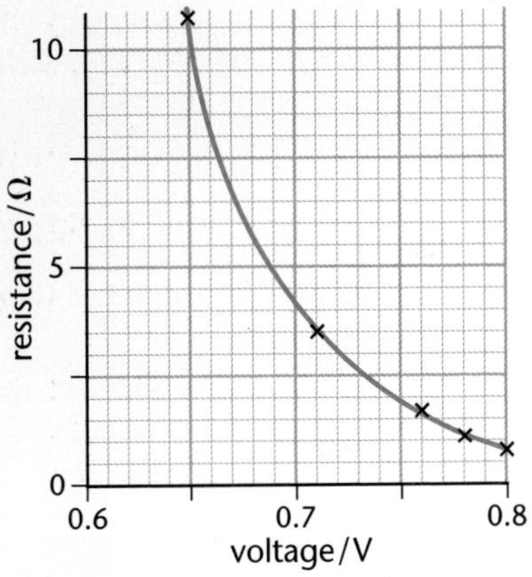

8 (a) $v = I/nAe$ [1] $= 5$ A $\div$ $(8.0 \times 10^{28}$ m$^{-3} \times 1.25 \times 10^{-6}$ m$^2 \times 1.60 \times 10^{-19}$ C) [1] $= 3.1 \times 10^{-4}$ m s^{-1} [1].

 (b) The cross-sectional area of the tungsten is smaller than that of the copper [1] and the concentration of free electrons is less [1].

 (c) Heating is due to collisions between the free electrons and the metal ions [1]. The collisions are more frequent in the tungsten than in the copper [1].

9 (a) Effective resistance of the two parallel resistors $= 4$ Ω [1].
   Circuit current, $I = V/R = 3.0$ V $\div$ 7.5 Ω [1] $= 0.40$ A [1].

 (b) $V = IR = 0.40$ A $\times$ 7.0 Ω [1] $= 2.8$ V [1].

Chapter 3 Particle physics

1 (a) 295 K [1].

 (b) $n = pV/RT$ [1] $= 1.20 \times 10^5$ Pa $\times 1.50 \times 10^{-4}$ m$^3 \div (8.3$ J mol$^{-1} \times 295$ K) [1] $= 7.4 \times 10^{-3}$ [1].

 (c) $T_2 = P_2T_1/P_1$ [1] $= 2.00 \times 10^5$ Pa $\times 295$ K $\div 1.20 \times 10^5$ Pa [1] $= 492$ K [1].

 (d) (i) $\frac{1}{2}m <c^2> = 3/2$ kT [1] $= 1.5 \times 1.38 \times 10^{-23}$ J K$^{-1} \times 295$ K [1] $= 6.11 \times 10^{-21}$ J [1]

   (ii) 590 K [1] since temperature is proportional to the mean kinetic energy of the particles [1].

2 Energy released when steam condenses to water at 100°C $= ml = 1.8$ kg $\times 2.25 \times 10^6$ J kg$^{-1} = 4.05 \times 10^6$ J [1]. Energy released when water cools from 100°C to 85°C $= mc\Delta\theta = 1.8$ kg $\times 4.20 \times 10^3$ J kg^{-1} K$^{-1} \times 15$ K $= 1.13 \times 10^5$ J [1]. Total energy released $= 4.16 \times 10^6$ J [1].

3 (a) (i) A temperature difference of 10 K can be measured more precisely than one of 1 K [1]

   (ii) There is less energy loss to the surroundings [1]

   (iii) The energy supplied heats the block and replaces the energy lost to the surroundings [1] so this should give a higher value [1]

   (iv) Insulate the block to reduce energy losses [1].

 (b) $c = E/m\Delta\theta$ [1] $= 24$ W $\times 350$ s $\div (0.80$ kg $\times 14$ K) [1] $= 750$ J kg^{-1} K^{-1} [1].

4 (a) ΔU is the increase in internal energy [1]
   Q is the heat supplied to the gas [1]
   W is the work done on the gas [1].

 (b) The temperature drops [1] as the heat removed from the gas is greater than the work done on it [1].

5 (a) Efficiency $=$ useful power output $\div$ power input $= 2$ kW $\div 8$ kW [1] $= 0.25$ [1].

 (b) $(T_H - T_c) \div T_H = (1023$ K $- 423$ K$) \div 1023$ K [1] $= 0.59$ [1].

 (c) Some energy is wasted as heat [1].

6 (a) (i) Particles that approach a nucleus are deflected away from it [1]. So the nucleus must have the same sign of charge as the alpha particles [1]

   (ii) Many particles pass through undeviated [1].

 (b) (i) High energy electrons [1] are used to penetrate nucleons [1]

   (ii) Electrons are not repelled by the nucleus [1] and they are not affected by the strong nuclear force [1]

   (iii) The proton and neutron are made up of other particles [1]

   (iv) The scattering of the electrons shows that there are dense regions of charge within a nucleon [1]

   (v) A proton is uud [1]; a neutron is udd [1].

7 (a) 6.02×10^{23} mol$^{-1} \times 1/12$ mol [1] $= 5.02 \times 10^{22}$ atoms [1].

 (b) (i) No [1] since 99% of carbon is carbon-12 [1]

   (ii) 5×10^{10} [1].

 (c) (i) $\lambda = 0.25 \div N$ [1] $= 0.25 \div 5 \times 10^{10} = 5 \times 10^{-12}$ s^{-1} [1]

   (ii) $t_{1/2} = 0.69 \div \lambda$ [1] $= 0.69 \div 5 \times 10^{-12}$ s$^{-1} = 1.38 \times 10^{11}$ s [1].

8 (a) If the neutron–proton ratio is less than 1 the nucleus undergoes β+ decay [1]; if the neutron–proton ratio is greater than 1 the nucleus decays by β- emission [1].

(b) When $^{13}_{6}C$ decays, the neutron–proton ratio changes from 1.17 [1] to 0.86, which is closer to 1 [1]

(c) It becomes greater than 1 [1] and reaches a value of 1.5 for massive nuclei [1].

9 (a) A meson consists of a quark and an antiquark [1]. A baryon is made up of three quarks [1].

(b) π^0 [1].

Chapter 4 Electromagnetic radiation

1 (a) In a transverse wave the vibrations are at right angles to the direction of wave travel [1].
Examples include any electromagnetic wave, e.g. radio, light [1].
In a longitudinal wave the vibrations are parallel to the direction of wave travel [1].
Examples include sound or any other compression wave [1].

(b) (i) In an unpolarised wave the vibrations are in all directions perpendicular to the direction of travel [1]. In a polarised wave the vibrations are in one direction only [1].

(ii) Sound is a longitudinal wave; there are no vibrations perpendicular to the direction of travel [1].

2 (a) The incident light, refracted light and normal are all in the same plane [1].
$\sin i \div \sin r = $ constant [1].

(b) (i) $\lambda = v \div f$ [1] $= 3.00 \times 10^8$ m s^{-1} $\div 6.00 \times 10^{14}$ Hz [1] $= 5.00 \times 10^{-7}$ m [1]

(ii) $n = c_{air} \div c_{water} = 3.00 \times 10^8$ m s^{-1} $\div 2.25 \times 10^8$ m s^{-1} [1] $= 1.33$ [1].

(iii) $\sin r = \sin i \div n = \sin 63° \div 1.33$ [1] $r = \sin^{-1} 0.668 = 42°$ [1].

3 (a) (i) 1.2 m [1]

(ii) $f = 1/T = 1 \div 2.50 \times 10^{-4}$ s [1] $= 4000$ Hz [1]

(iii) $v = f\lambda$ [1] $= 4000$ Hz $\times 1.2$ m [1] $= 4800$ m s^{-1} [1].

(b) (i) The wave profile does not move along the rope [1]

(ii) By superposition [1] of a wave and its reflection [1].

(c) (i) 0° [1]

(ii) 180° [1].

4 (a) (i) The waves are in phase [1] and interfere constructively [1]

(ii) The waves are out of phase [1] and interfere destructively [1].

b) (i) $\lambda = v \div f$ [1] $= 330$ m s^{-1} $\div 2000$ Hz $= 0.165$ m [1]

(ii) Separation of maxima, $x = \lambda D/a = 0.165$ m $\times 5.0$ m $\div 0.75$ m $= 1.10$ m [1].
CE is half this distance $= 0.55$ m [1].

5 (a) $_g n_p = {_a}n_p \div {_a}n_g$ [1] $= 1.24 \div 1.58 = 0.785$ [1].

(b) $\sin C = 0.785$ [1] $C = \sin^{-1} 0.785 = 52°$ [1].

(c) Light has further to travel along path B than along path A [1]. This causes a pulse to be elongated [1].

6 (a) (i) Emission of electrons [1] when a metal surface is illuminated with electromagnetic radiation [1]

(ii) A quantum of electromagnetic radiation [1] of a particular frequency [1]

(iii) The minimum energy [1] an electron needs to escape from a metal [1].

(b) (i) $f_0 = \phi \div h$ [1] $= 5.05$ eV $\times 1.60 \times 10^{-19}$ C $\div 6.63 \times 10^{-34}$ J s $= 1.22 \times 10^{15}$ Hz [1]

(ii) $(\frac{1}{2}mv^2)_{max} = hf - \phi$ [1] $= 6.63 \times 10^{-34}$ J s $\times 1.80 \times 10^{15}$ Hz $- (5.05$ eV $\times 1.60 \times 10^{-19}$ C) [1] $= 3.85 \times 10^{-19}$ J [1].

Index